Études d'archéologie 20

FORGING VALUES

METALS TECHNOLOGIES IN THE AEGEAN AND BEYOND FROM THE 4th TO THE 1st MILLENNIUM BCE

Nerantzis Nerantzis (ed.)

Bruxelles
CReA-Patrimoine
2023

Éditeur
CReA-Patrimoine

Université libre de Bruxelles
50, av. F.D. Roosevelt | CP 133/01
B-1050 Bruxelles
crea.centresphisoc.ulb.be

ISBN : 978-2-9602029-7-7
Impression : Snel
Mise en page : Nathalie Bloch (CReA-Patrimoine)

Couverture
Londres, British Museum 1846.6-2945, Keyside Painter, artisans blacksmiths (drawing by N. Bloch (CReA-Patrimoine), after W. Nicoll, *A Catalogue of the Greek and Etruscan Vases in the British Museum*, Londres 1851, n° 668, pl. XXIII.)

Études d'archéologie 20

FORGING VALUES

METALS TECHNOLOGIES
IN THE AEGEAN AND BEYOND
FROM THE 4th TO THE 1st MILLENNIUM BCE

CONTENTS

PREFACE

The significance of early metallurgy, its organisational complexity and the associated far-reaching networks for exchanging its valuable products have been central themes in archaeological research since long ago. Regions rich in mineral deposits became a focus for researchers who study mining and metallurgical processes by employing various interdisciplinary methodologies and interpretative tools. In the context of previous research frameworks stratified societies, states or kingdoms have been often described as important consumers stimulating demand and circulation of metals as either raw materials or finished products. Crafts production manifested in the rural mining regions could have been somewhat different from the urbanised areas where crafting was often dependent on specialised labour under administrative control and such issues are deemed crucial in the study of ancient metal "industries". In this context, itinerant craftsmen are often seen as crucial agents of exchange and spread of technological knowledge crossing cultural boundaries.

The north Aegean represents one such region where natural resources and mobility of populations have apparently favoured a long-lasting metal producing tradition emerging in the Late Neolithic and growing in significance particularly during the Hellenistic and Roman periods. However, interpretations based on neo-evolutionary ideas describing "progress" as a slow transformative process from simple to gradually more complex technological stages, do not seem to be supported by actual archaeological evidence. Moreover, such models had direct influence on the way we understand social organisation following linear progressions from simple to more complex forms. Since technology could be described as an embedded set of social acts, entangled with culturally conditioned practice, the study of particular contexts becomes important in offering concrete examples on how production and consumption of metals are socially determined. In the north Aegean, at least during the early stages, metals do not seem to signify social order while metallurgical technology operates in contexts upon which no apparent control was exercised as discussed in the first chapter of this book. The geographical location of Greek Eastern Macedonia and Thrace, connecting the Aegean with the Balkans and Anatolia, the rich mineral deposits and vibrant interactions of local communities with incoming populations played a crucial role in technological exchange. Production of base and precious metals seems to have been an identity forming element for this region as hinted in the recurring references to its rich metalworking tradition by several ancient authors. The Marie Skłodowska-Curie funded project *Metals Technology in North Aegean Societies* (*ME.Tech.NAS.*) focused on metals production and consumption in this region while exploring the possible connections with neighbouring or more distant lands. Attempting at a more thorough understanding of complex processes involving the maintenance of supra-regional networks, it was decided to posit metals production and evidence for connectivity from the study region into a broader geographical perspective. Keeping such considerations in mind the International Symposium "*Forging Values*" was held at Brussels on the 15th and 16th of April 2021 as an online event. It focused on mining, ore processing, metallurgy, trade and deposition of metals from the Late Neolithic/Early Bronze Age to the Hellenistic period in Greece and surrounding regions aiming to attempt a vivid demonstration of current and ongoing research. Participants to the conference addressed important issues such as the production and trade of copper and silver during the Bronze Age, the emergence and adoption of iron technology, the importance of gold and silver used as pre-monetary media for transactions and the large-scale extraction and purification of precious metals during Classical antiquity.

This volume presents a selection of papers covering part of the presented research topics and its scope is to contribute to the field of interdisciplinary studies on ancient metallurgy (archaeometallurgy) with new results, methodological approaches and novel interpretative tools. The chapters that follow cover a wide chronological span that starts at the end of the Neolithic in the late 4th millennium BCE and ends up to Hellenistic times in the 4th c. BCE.

The first chapter presents the early steps of metal production in north-eastern Greece starting with the earliest example of copper smelting from Late Neolithic Promachon-Topolnica, and followed by metalworking evidence at Sitagroi and Dikili Tash. Discussion is also focused on the island of Thasos with its rich mineral deposits that yielded the earliest evidence for silver extraction in the North Aegean. Georgakopoulou presents a thorough study of archaeometallurgical assemblages from several seasons of survey and excavation projects at Dhaskalio and Kavos on Keros in the Cyclades. Important new data are presented, contributing in this way to the growing body of evidence on EBA Cycladic metallurgy for the production of copper, lead and silver. Oudbashi and Mehofer present an overview of analytical results on a large number of copper-based artefacts to showcase the emergence and development of tin bronze technology during the Bronze Age in Iran. In this context the transition from arsenical copper to tin bronze metallurgy is thoroughly discussed. Stos-Gale attempts a reassessment of the available lead isotope data on metal artefacts from sites in the Aegean, dating between the LN and LBA, in view of the recent discoveries of evidence for early metallurgy of copper, silver and lead in northern Greece. Heymans and Verdan discuss the role of natural gold sources and their exploitation in the changing societies of the northern Aegean between the end of the Late Bronze Age and the archaic period. Their approach follows a historical/cultural perspective to help untangle the ways by which metal resources helped shape new values and how changing patterns of exploitation and use, as well as the human interactions involved, can be tied to changes in history and culture. Nodin focuses on a large corpus of stone tools used in ore processing from the ancient city of Thasos. The article presents important findings on typological associations, and the evolution of techniques for crushing and grinding mineral ores as revealed through the use of such toolsets. Nerantzis and co-authors present an investigation of the contexts that yielded a significant amount of metallurgical debris in the city of Pistyros during the ongoing systematic excavation project. The overwhelming volume and wide distribution of finds related to metallurgy across the excavated Hellenistic layers, and some preliminary analytical results are used in a first attempt to interpret the evidence. Photos-Jones offers an overview of terms encountered in ancient literary sources to denote metallurgical residues of different sorts. She argues that examining how ancient craftspeople called their by-products is fundamental in attempting to reconstruct the complex *chaînes opératoires* for producing metals and eventually pharmacological preparations. At the same time the article offers a critical re-examination of long enduring misconceptions founded on the misuse of modern industrial terminology in an attempt to understand ancient practices. Chalazonitis and Malamidou cover the topic of consumption of metals in burial contexts with the example of two graves in the Strymon River plain. The important issue of communicating social roles or "elite" identities through preferential consumption in funerary assemblages offers an interesting interpretative tool in understanding social organisation through material culture.

Nerantzis Nerantzis
2022

FOREWORD

The Proceedings of the symposium "Forging Values" edited by Dr Nerantzis Nerantzis is the 10th publication in the field of Greek archaeology in the series of "Etudes d'Archéologie" of the CReA-Patrimoine and the 1st volume that concerns ancient Greek metallurgy in the same collection. These figures reflect the important role played by Dr Nerantzis together with Dr Ioannis (Yangos) Chalazonitis, in the development of this area of expertise at the CReA-Patrimoine.

From the very beginning, the studies in Greek archaeology at the CReA-Patrimoine focused on social and economic aspects of craft productions, especially pottery, in ancient societies, including a large spectrum of productions (fine, cooking and coarse wares) along with wide geographical areas (Egypt, Near East, Greece, Roman) and chronological periods (from the Archaic to the Medieval period). The present volume shares this interest in cultural and economic issues of craft productions, but it expands the discussion to the metal production and consumption in the North Aegean during a wide chronological span, from the Late Neolithic to the Hellenistic periods. It further lays emphasis on technological questions about craft organisation, trade and distribution networks, with a specific interest in ancient metal technology, providing us with a real interdisciplinary approach that discusses questions about mining, ore processing, the emergence of new technologies and their impact on trade and cultural exchanges.

Nerantzis Nerantzis joined the CReA-Patrimoine in the frame of a Marie Sklodowska-Curie postdoctoral grant, funded by the European Union's Horizon 2020 research and innovation programme. During his stay in Brussels, Nerantzis successfully organized one workshop and an international symposium, both events resulting from several questions he was discussing in his own research on the role played by the development of ancient technological knowledge and the spread of metallurgical traditions. This volume that resulted from the international symposium brings together the leading scholars in the field and will certainly become a reference on the subject.

I would like to use the opportunity of this foreword to warmly thank Nerantzis for his fruitful collaboration at the CReA-Patrimoine and for the numerous exchanges and discussions with all the research fellows. Many thanks are also due to the successive Directors of the Centre, Sebastien Clerbois and François Blary, for providing us with an unfailing assistance to all initiatives related to ancient Greek archaeology. I am also very grateful to my old friend and colleague, archaeologist-infographist, Nathalie Bloch, who prepared the manuscript for publication with her well acknowledged talent and care, despite the difficult circumstances. Last but not least, I would like to thank the Université libre de Bruxelles (ULB) for its continuous support to the development of our studies in the field.

Athena Tsingarida

ACKNOWLEDGMENTS

The Symposium "Forging Values" was conducted in the framework of a research project that received funding from the European Union's Horizon 2020 research and innovation programme under the Marie Sklodowska-Curie grant agreement No 840894 – Metals Technology in North Aegean Societies (ME.Tech.NAS). The editor wishes to thank warmly Prof. Athena Tsingarida for her insightful supervision, coordination of research and the excellent cooperation. Thanks are also due to the following colleagues: S. Nodin, S. Papadopoulos, E. Photos-Jones, Y. Bassiakos, E. Filippaki, Ch. Koukouli-Chrysanthaki, D. Matsas, C. Kallintzi, D. Malamidou, S. Dadaki, M. Tasaklaki, D. Terzopoulou, N. Arrington, D. Tsiafaki, V. Evangelidis, N. Epitropou, D. Vivier, K. Neeft, V. Saripanidi, I. Chalazonitis, T. Manolova, C. Judson, A. Attout and N. Bloch.

The volume is dedicated to the memory of our dear colleague Myrto Georgakopoulou.

1. INTRODUCTION
METALS TECHNOLOGY IN NORTH AEGEAN SOCIETIES: THE EARLY STEPS

Nerantzis Nerantzis

This introduction sets the scene to the early steps of prehistoric metallurgy in the north Aegean drawing on recent interdisciplinary research and attempts to address the role that metals technology played, whilst being embedded in early forms of socio-political organisation in this region. The north Aegean is often portrayed as a periphery of the Greek world that was drastically influenced by socio-political developments taking place in southern Greece but at the same time shaped by cultural affinities and links oriented towards the north. Although geographically being an extension of the Balkan hinterland that witnessed an early rise of stratified societies in the mid-5th millennium using copper and gold in conspicuous burial customs,[1] the trinket output in metals from the north Aegean during the same period is somehow intriguing. [2] It is conventionally seen as a sort of buffer zone during the Late Bronze Age between the Mycenaean palatial centres to the south, the Balkan hinterland and the western Anatolian urban centres, displaying less noticeable signs for social hierarchies controlling trade and influx of goods and ideas.[3] In explanations based on centre-periphery dynamics or World Systems Theory, exchange or trade is seen to be conducted along core centres, peripheries and intervening transition zones, with the northern Aegean fitting the latter domain.[4] Although such models seem to have interpretative qualities at macro-scale levels, they often describe agency driven by the central places, while peripheries at their micro-regional level are seen as passive or impervious to cultural complexity.[5]

Such perspectives have been partly changed due to recent excavation projects that revealed important Bronze Age settlements at Limenaria and Skala Sotiros on Thasos[6] and Mikro Vouni on Samothrace[7] which maintained active links to the southern and eastern centres long before the LBA (Fig. 1). At the EBA II fortified site of Skala Sotiros anthropomorphic stelae showing parallels to Troy and Soufli Magoula in Larisa were found incorporated to the fortification wall.[8] Further, the finding at Mikro Vouni of roundels and noduli bearing seal impressions of Linear A hieroglyphic signs along with Minoan or Minoanizing pottery are indicative to such external links.[9] These findings confirmed that the north Aegean islands and to some extent the opposite coast to the north participated to Bronze Age long distance networks of exchange. Due to the scarcity of detailed studies, a supposedly peripheral role in terms of cultural complexity has led to a skewed portrayal of the region as a cultural backwater, an idea that is gradually being revised with ongoing research.[10]

It is rather telling that the region's archaeological record features more prominently in academic literature from the Greek colonisation Phase of the Archaic period onward,[11] when it becomes relevant to important political events related to the rise of hegemonic powers in the Aegean, namely the Athenian state and the Macedonian kingdom. In this context the natural resources of the north Aegean, particularly precious metals, have been often seen as a motive behind colonial endeavors, establishment of spheres of influence or an arena of armed conflict for the securing of access to silver and gold mines.[12] This approach although partially efficient at offering

1 Ivanov and Avramova 2000; Leusch *et al.* 2017.

2 Zachos 2010, 81-82.

3 Andreou 2010; Chapman 2014, 34.

4 Bintliff 1997; Kristiansen 1998.

5 Hall *et al.* 2010, 58; Kohl 2011.

6 Papadopoulos and Malamidou 2012; Koukouli-Chrysanthaki and Papadopoulos 2016.

7 Matsas 1991; 1995.

8 Koukouli-Chrysanthaki 1988, 396.

9 Matsas 2010.

10 Matsas *et al.* 2011; Girella and Pavúk 2015.

11 Tiverios 2008; Baralis 2009.

12 Zannis 2014.

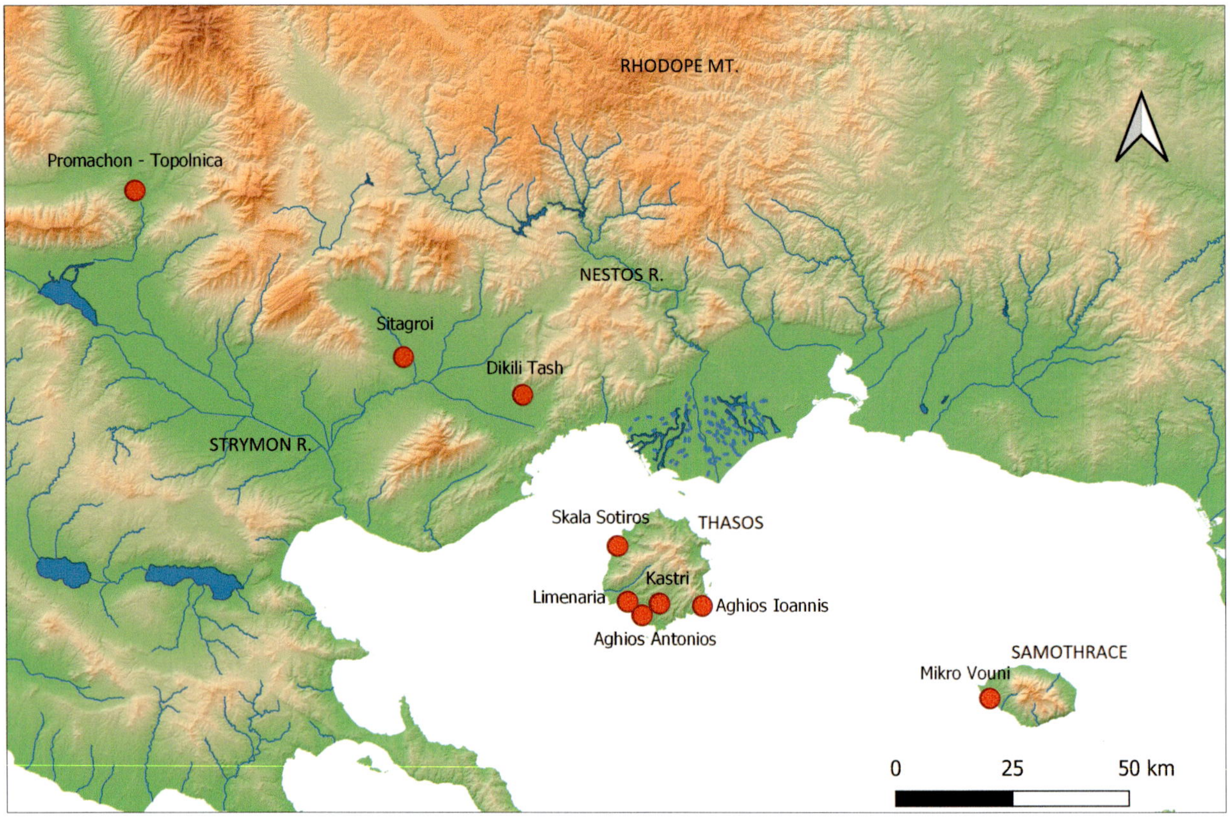

Fig. 1. Map of north Aegean showing sites with evidence for early metallurgy (designed by V. Evangelidis).

potential explanations on macro-scale agency fails to interpret the role of other players such as miners, craftsmen and private entrepreneurs in the transfer of technological knowledge, spread of innovations and investment in the running of such industries.

The rich mineral resources of the region are well known and represent the largest and more densely concentrated metal-bearing deposits compared to any other region in Greece. Although featuring prominently in numerous debates on the provenance of metals, their prehistoric exploration and extraction of metals has not attracted adequate interdisciplinary research.[13] Similarly, the study of mining and metal production of antiquity through geological prospection combined with limited archaeological fieldwork has been intermittent since the 1970s and thus an overall synthesis on the subject is lacking.[14] The major aim of the research project *Metals Technology in North Aegean Societies* (*ME.Tech.NAS.*) has been to investigate the techniques for processing mineral raw materials and technologies involved in metal production across this region from the Late Neolithic up to the Hellenistic period. This was attempted through targeted case studies seeking to explore specific research questions, namely the early emergence of copper and silver metallurgy, the adoption of iron technology, and large-scale precious metals extraction of later periods. Such an approach is useful in providing quantifiable data on technical intricacies in metal production and possible transfers in technological information that would help in approaching wider socio-cultural syntheses for this region. This introduction focuses on early copper and silver production from the Final Neolithic to the end of the Bronze Age.

Recent studies have confirmed that the inception of metallurgy in the Balkans, around the end of the 6th millennium (c. 5200-5000 BC), emerged on gradual steps of trial and error and long experimentation that led to the smelting of copper ores in ceramic crucibles, covered with charcoal, to which air was introduced through blow pipes.[15] The earliest finds

13 Mcgeehan-Liritzis 1983.

14 Wagner and Weisgerber 1988; Photos *et al.* 1989; Koukouli-chrysanthaki 1990; Vavelidis *et al.* 1997.

15 Thornton and Roberts 2009; Roberts *et al.* 2009; Radivojević *et al.* 2010; Radivojević *et al.* 2021.

(c. 5000 and 4600 BC) related to the smelting of copper ores derive from settlements of the Vinča culture namely Belovode, Vinča-Belo Brdo, Plocnik, Gorja Tuzla and Selevac in Serbia.[16] These finds are in form of copper slag and slagged sherds found at both Belovode and Gornja Tuzla, which suggest the presence of a hole-in-the-ground installation lined with broken pottery.[17] By the mid to late 5th millennium (c. 4600-4100 BC) copper smelting is evidenced at Dolnoslav, Chatalka and Akladi Cheiri in Bulgaria.[18] All three sites yielded crucibles, amongst other finds, while at Akladi Cheiri there is evidence for hole-in-the-ground smelting installations lined with pottery sherds, similar to the earlier examples from Serbia.[19]

These technological innovations seem to have spread to the northern Greek territories that were sharing a whole set of cultural traits with the lower Danube and Struma valley in terms of architecture, figurative art, pottery and lithic technologies.[20] The period from 4800 to 4300 BC, during which copper and gold finds increase, witnessed the rise of important cultural groups, namely Vinča C and D in Serbia and western Rumania, Kodzadermen and Karanovo in Bulgaria and Gumelniţa in Rumania.[21] In northeastern Greece excavations at the settlements of Sitagroi,[22] Dikili Tash,[23] Promachon-Topolnica[24] and Dimitra,[25] Diomedeia,[26] Paradimi[27] and Makri[28] have yielded a wealth of archaeological finds displaying explicit cultural affinities to their hinterland Balkan counterparts.

The earliest archaeological evidence for the primary processing of copper ore in northern Greece was recovered at the site of Promachon-Topolnica, in a layer dating between 5070 and 4700 BC.[29] The settlement is situated on the Greek/Bulgarian border and excavations revealed a flat, extended habitation surface that spread on both sides of the frontier. Fragments of crucibles and malachite lumps found in a workspace consisting of clay-lined cavities, where high temperature reactions took place, represent proof of copper production practices. In fact, the available absolute dates place this assemblage, along with contemporary evidence from Belovode in Serbia,[30] among the earliest copper production finds recovered so far from a European context. The ongoing archaeometallurgical study at N.C.S.R. "Demokritos" including in situ pXRF measurements, sampling and compositional analysis of burnt soil from the cavities' interior has so far confirmed that secondary copper minerals were thermally treated within these features. Careful examination of the sampled soils under the stereoscope confirmed the presence of minute malachite grains while compositional analysis revealed increased copper contents in some burnt soil samples. It could be suggested that the absence of slag from the crucible's interior is due to the use of pure malachite, rich in copper content, which does not leave slag during smelting. The quantities of produced metal were most probably restricted, while the finding of only minute copper fragments in form of beads further corroborates the trinket nature of the output.

The settlements of Sitagroi and Dikili Tash in the plain of Drama have yielded some of the earliest copper and gold objects. At Sitagroi apart from several copper utensils mainly awls, hooks and pins evidence for local manufacture mainly clay crucibles for the melting of copper that date to the middle of the 5th millennium were found.[31] The lead isotope analyses suggested that copper was reaching the site from a variety of sources with the earlier artefacts dating to Phase III assigned to sources in the north, that is the Balkan hinterland.[32] A southern Aegean source is suggested for the artefacts belonging to the Bronze Age (Phases IV and V). The shifting of procurement sources hinted in the lead isotope data is in accordance with changes in composition from pure copper to tin and leaded bronze which might reflect the establishment of new networks of exchange by the EBA directed towards the south and east.

At Dikili Tash recent excavation has yielded a clay mould decorated with incisions dating to the LN II,[33] several copper alloy finds and pieces of gold sheet

16 Radivojević and Roberts 2021, 19.

17 Radivojević and Roberts 2021, 20.

18 Todorova 1999, 242; Leshtakov 2013; Rehren *et al.* 2020.

19 Ryndina *et al.* 1999; Rehren *et al.* 2020.

20 Aslanis 1988; Malamidou *et al.* 2006.

21 Anthony 2010, 35.

22 Renfrew *et al.* 1986; Elster and Renfrew 2003.

23 Treuil 1992; Darque *et al.* 2007.

24 Koukouli-Chrysanthaki *et al.* 2007.

25 Grammenos 1997.

26 Kallintzi and Papadopoulos 2010.

27 Bakalakis and Sakelariou 1981.

28 Efstratiou and Kallintzi 1997.

29 Koukouli-Chrysanthaki and Bassiakos 2002, 194.

30 Radivojević *et al.* 2010, 2779.

31 Renfrew and Slater 2003, 303.

32 Stos-Gale 2003, 325-329.

33 Darque *et al.* 2009, 533.

rolled around stone beads as well as a gold spiral dating between 4400 and 4200 BC.[34] As the excavators have pointed out although this latter type of objects is relatively common in the Aegean-Balkan area of the 5th millennium BC such assemblages rarely appear outside a burial context.[35]

The case of Thasos stands out since the island is rich in complex mineral deposits and their early extraction is reflected in metallurgical evidence from two settlements, namely Limenaria and Aghios Antonios situated on the southwestern coast, not far from copper and argentiferous lead-bearing mineral deposits.[36] Metalworking finds were also found at Skala Sotiros along with a few copper alloy artefacts and a gold perforated discoid ornament (pectoral), the sole example from Early Bronze Age Thasos. It was most probably fitted on clothing and has parallels with contemporary gold jewellery from Troy and Poliochni.[37]

From the settlement of Limenaria, derive the earliest residues of silver extraction from the entire north Aegean region. These are in form of bowl-shaped litharge-resembling fragments deriving from a cupellation process and a sliver pin, all recovered from a layer dating to the early 4th millennium (c. 3900-3700 BC).[38] These important finds have been the subject of interdisciplinary investigation at N.C.S.R. "Demokritos". The chemical composition of litharge is PbO, which forms through metallurgy when molten argentiferous lead is subjected to oxidizing conditions within suitable vessels or clay-lined pits in order to separate the two metals. Analyses of the litharge-resembling pieces from Limenaria and similar finds from contemporary sites in Attica, suggest that these do not consist of pure lead oxide, but they include increased contents of siliceous and calcareous compounds.[39] These fragments should be thus characterized as pieces of litharge-impregnated cupellation hearth lining. This means that while the lead was in liquid state it was mostly absorbed by the ceramic vessel within which cupellation was taking place leaving silver to float on top and thus to be separated.

Similar finds in much greater quantities derive from several sites in Attica mainly Koropi, Lambrika, Merenda and Zapani[40] and also from Akrotiraki in southern Siphnos[41] showing that the major silver-bearing deposits in the Aegean were already utilized at the beginning of the Early Bronze Age. It is interesting to note that bowl and oval shaped litharge pieces of similar typologies were found at Fatmali Kalecik in eastern Anatolia,[42] at Arisman[43] and Tepe Sialk[44] in Iran dating to the mid-4th millennium and at Habuba Kabira[45] in northern Syria of the second half of the 4th millennium. The roughly contemporary emergence of this technology across the Near East and the Aegean may attest to active links and inter-regional exchange of technological information crossing cultural boundaries. Notably the north Aegean seems incorporated in these innovative technological networks as the finds from Limenaria attest.

The earliest evidence for copper metallurgy on Thasos derives from the same settlement in form of slag pieces, crucible fragments, and copper alloy objects found in Early Bronze Age I layers, dating around 3000 BC.[46] More findings related to copper production mainly slag, clay lining and a mould have come to light at the settlement of Aghios Antonios, near Potos dating between c. 2900 and 2400 BC. At Skala Sotiros, an EBA II settlement on the NW coast, stone tools for ore processing and metalworking utensils such as a double mould and tuyères were retrieved.

Based on instrumental analysis of the finds from Limenaria and Aghios Antonios employing optical microscopy, XRF and SEM/EDS it has been shown that copper extraction was achieved through smelting of secondary cupriferous ores in which other gangue and metallic minerals co-existed[47]. The presence of arsenic within a few copper prills in contents between 2 and 4% might suggest that at least some of the slags derive either from the smelting of copper ores rich in arsenic or the co-smelting of arsenic-bearing ores along with copper minerals.

Besides the study of the technical intricacies involved

34 Darque *et al.* 2012.

35 Darque *et al.* 2012, 757.

36 Nerantzis *et al.* 2016.

37 Koukouli-Chrysanthaki 1988, 428.

38 Papadopoulos 2008, 69.

39 Bassiakos *et al.* 2019, 2752.

40 Georgakopoulou *et al.* 2020, 185-186.

41 Papadopoulou 2011.

42 Hess *et al.* 1998.

43 Helwing 2014, 415.

44 Pernicka 2004.

45 Pernicka *et al.* 1998.

46 Bassiakos *et al.* 2019, 2747-2751.

47 Nerantzis *et al.* 2016, 576-577.

in metalworking it is equally crucial to examine the social organization and cultural context within which the prehistoric communities of the region were utilizing mineral resources. The abundant use of bright colours in pottery with highly polished surfaces as well as the use of spondylus shells and beads of colourful minerals for display are typical of the 5th millennium Balkan cultural "package'.[48] Within this aesthetic milieu the introduction of copper and gold ornaments communicated powerful messages about social rank, reflecting in this way the articulation of a progressively more complex social order.[49]

The distribution of copper and gold objects in Balkan settlements and cemeteries clearly shows that they were found in a minority of houses and graves. Thus, the assumption that copper and gold objects were markers of status, prestige, or wealth is not unreasonable.[50] It seems that control of precious resources and access to specialist metallurgical knowledge could have functioned as elements for constructing the arena within which social inequality was expressed. It is not certain how the communities of the north Aegean territories responded to such changes in the construction of social hierarchies for which evidence is more prominent further north. Since no burials of the Late Neolithic have been excavated in northeastern Greece that would account for the presence of social groups exerting control on resources or having differential access to precious materials it is difficult to draw any safe conclusions.

So far, there are no convincing signs to account for the existence of particular social groups controlling access to metal-bearing resources during the Final Neolithic. The earlier crucible fragments from Promachon-Topolnica found within domestic contexts reveal that its inhabitants had access to the raw material and the technical know-how to process and transform ores into usable items. Therefore, a domestic milieu characterised by cooking, spinning, weaving, firing pottery or stone and bone tool making with no apparent ritual connotations represents the arena within which early metalworking appears to be embedded. Access to the knowledge of copper smelting does not seem to operate within a secluded context upon which control could be exerted. Yet the scarcity of information on hierarchical social ordering in the examined cases, or rather the complete lack of it, does not allow for more nuanced interpretations on consumption patterns. The silver pin and litharge from Limenaria hint to the early and successful attempts to extract lead and silver using the local argentiferous minerals. The production is overtly of a small-scale, domestic character but nevertheless involves sophisticated technical knowledge that was available in contemporary metallurgical centres in the Aegean and farther to the east.[51]

Copper metallurgy of the Early Bronze Age appears well embedded within other domestic activities and although deposition of metal objects in burials remains elusive some clues on identity formation through the consumption of metals may be found at Skala Sotiros. There, during construction of the fortification wall dating to the EBA II period, several anthropomorphic stelae had been incorporated as building material which should belong to an earlier Phase of the EBA I.[52] These are marble fragments of stylized human figures and a slab-shaped stele of the menhir type. It is important to note that these stelae were interpreted as representations of hunters, warriors or venerated ancestors considering the belts and weapons they were portrayed with. Interestingly one of these figures is depicted holding a metal dagger in one hand while an axe or hatchet is attached to the belt. Such elements represent manifestations of rank and social esteem as the use of metal weapons in ceremonial display has been associated to chiefs or powerful individuals in numerous Bronze Age contexts across Europe, the Mediterranean and the Near East. The access to metals for social display and the control of their production and circulation has been a crucial factor in the dynamic process that gave rise to wealthy elites in the Bronze Age Aegean and elsewhere. Although the evidence so far is limited for understanding the role of special groups or elites in controlling resources and encouraging craft specialization in EBA Thasos, the stelae provide further indirect proof for metal circulation and consumption on the island during this period. The finding of the gold pectoral at the same site might also underline the same pattern in using precious materials for social display.

Crucial links seem to have developed by the MBA between the northern islands and the palatial centres to the south[53] possibly associated to an impetus for the search of metals in the north as reflected in the Minoan presence at Mikro Vouni on Samothrace[54]

48 Chapman 2003; Gaydarska and Chapman 2008.

49 Leusch *et al.* 2017.

50 Porčić 2019, 356.

51 Helwing 2014.

52 Koukouli-Chrysanthaki 1988, 396.

53 Girella and Pavúk 2016.

54 Matsas 1995.

and Koukonisi on Lemnos.[55] The coexistence of metallurgical evidence in association with Minoan or Minoanising material on both sites should not be the result of mere coincidence. The straight cylindrical tuyères from Koukonisi show direct parallels to Ayia Eirini on Kea and Poros Katsambas, Mochlos, Kommos and Palaikastro on Crete.[56] Likewise, the mould appears as a hybrid type combining a typical axe shape widespread in the Balkans and western Anatolia but the incised marks on its surface are common in south Aegean contexts.[57] Connectivity becomes more pronounced over the course of the Late Bronze Age as suggested by imported Mycenaean pottery from LH IIIB and LHIIIB-IIIC transition period and local imitations on Thasos[58] and some mainland sites. Contacts with the Mycenaean palatial centres brought a new stimulus for further economic exchange also linked to the technological sphere possibly related to the introduction of iron technology.

Tin bronze becomes more frequent in settlements and burial contexts while the two early iron knives found at the cemeteries of Kastri on Thasos were most probably imported from Mycenaean centres. Their appearance seems to coincide with Mycenaean trade as suggested by blue glass, amber beads and bronze knives of Mycenaean types found in the same cemeteries.[59] Previous explanations relied heavily on the hypothesis of a wholesale introduction of iron technology to the Aegean from Cyprus but more recently such interpretations have been revised. Iron objects were in fact produced in Mycenaean centres already during the Late Bronze Age, a tradition that continued well up to the end of the palatial period in the Peloponnese.[60] As far as the knives from Thasos are concerned these seem to follow a locally rooted Aegean tradition of object types.

We could therefore envisage the north Aegean during the Late Bronze Age as a metal producing as well as metals receiving territory located at the convergence of important trade routes for copper and possibly also silver and gold from the Balkan and Anatolian trade routes to the Aegean, as well as tin from further afield. Workshops and trading posts active in nodal points on the islands served as catalysts for the spreading of technological knowledge originating from several centres of well-established metallurgical traditions. Access to abundant and diverse metal bearing ores, the local tradition of copper and silver extraction and exchange with other centres as well as the early adoption of iron could have accelerated the rise of local elites. Metals played a crucial role in the succeeding Early Iron Age and Archaic periods as the region became gradually incorporated to the Greek world. After all, the establishment of numerous Greek *emporia* and *apoikiai* in Aegean Thrace was, to some extent, targeted to a more systematic exploitation of its rich mineral resources.

55 Boulotis 2009.

56 Boulotis 2009, 186.

57 Boulotis 2009, 203.

58 Koukouli-Chrysanthaki 1992.

59 Koukouli-Chrysanthaki1992, 822.

60 Pallermo 2018, 116.

Bibliography

S. Andreou, "Northern Aegean", in: E.H. Cline (ed.) *The Oxford Handbook of the Bronze Age Aegean*, Oxford-New York, 2010, 643-659.

D.W. Anthony, "The rise and fall of Old Europe", in: D.W. Anthony and J.Y. Chi (eds), *The Lost World of Old Europe, The Danube Valley, 5000-3500 BC*, Princeton University Press, Princeton and Oxford, 2010, 28-57.

I. Aslanis, "Η Νεολιθική και η Πρώιμη Εποχή του Χαλκού στην Αιγαιακή Θράκη" in: *Πρακτικά Συμποσίου "Η Ιστορική, αρχαιολογική και λαογραφική έρευνα για τη Θράκη"*, Θεσσαλονίκη, 1988, 139-158.

G. Bakalakis and A. Sakelariou, *Paradimi*, Heidelberger Akademie der Wissenschaften Monographien Bd. II, Mainz, 1981.

A. Baralis, "The Chora Formation of the Greek Cities of Aegean Thrace. Towards a Chronological Approach to the Colonization Process", in: P. Guldager Bilde, J. H. Petersen (eds), *Meetings of Cultures in the Black Sea Region: Between Conflict and Coexistence*. Aarhus, Aarhus University Press (Black Sea Studies 8), 2009, 101-130.

J. bintliff, "Regional Survey, Demography, and the Rise of Complex Societies in the Ancient Aegean: Core-Periphery, Neo-Malthusian, and Other Interpretive Models", *Journal of Field Archaeology*, Vol. 24, 1 (1997), 1-38.

Ch. Boulotis, "Koukonisi on Lemnos: reflections on the Minoan and Minoanising evidence", in: C.F. Macdonald, E. Hallager and W.D. Niemeier (eds), *The Minoans in the Central, Eastern and Northern Aegean: new evidence*, 8th edn. Monographs of the Danish Institute at Athens, Athens, 2009, 175-218.

J. Chapman, "Colour in Balkan prehistory (alternatives to the Berlin & Kay paradigm)", in: L. Nikolova (ed.), *Early Symbolic Systems for Communication in Southeast Europe*, BAR Intern. Series 1139, Archaeopress, Oxford, 2003, 31-56.

R. Chapman, "Scales, interaction and movement in later Mediterranean prehistory", in: S. Souvatzi and A. Hadji (eds), *Space and Time in Mediterranean Prehistory*. Routledge Studies in Archaeology. Routledge, New York and London, (2014) 32-48.

P. Darque, Ch. Koukouli-Chrysanthaki, D. Malamidou, R. Treuil and Z. Tsirtsoni, "Recent researches at the Neolithic settlement of Dikili Tash: an overview", in: H. Todorova, M. Stefanovich and G. Ivanov (eds), *Proceedings of the International Symposium "Strymon Praehistoricus*', Sofia, 2007, 247-256.

N. Efstratiou and C. Kallintzi, "Αρχαιολογικές έρευνες στη Μάκρη Έβρου. Εκτιμήσεις και προβλήματα", *Το Αρχαιολογικό Έργο στη Μακεδονία και τη Θράκη 10Β* (1996), 1997, 881-916.

E.S. Elster and C. Renfrew, *Prehistoric Sitagroi: Excavations in Northeast Greece, 1968-70, Volume 2, The Final Report*, Monumenta Archaelogica 20, Los Angeles, 2014.

B. Gaydarska and J. Chapman, "The aesthetics of colour and brilliance - or why were prehistoric persons interested in rocks, minerals, clays and pigments?", in: R. I. Kostov, B. Gaydarska and M. Gurova (eds), *Geoarchaeology and Archaeomineralogy, Proceedings of the International Conference*, Publishing House "St. Ivan Rilski", Sofia, 2008, 63-66.

M. Georgakopoulou, K. Douni, M. Ginalas, O. Kakavogianni and Y. Bassiakos, "Recent Finds from Final Neolithic and Early Bronze Age Silver Production Sites in Southeastern Attica", in: N. Papadimitriou, J.C. Wright, S. Fachard, N. Polychronakoy-Sgouritsa and E. Andrikou (eds), *Athens and Attica in Prehistory, Proceedings of the International Conference*, Oxford, Archaeopress Publishing, 2020, 185-192.

L. Girella and P. Pavúk, "Minoanisation, Acculturation, Hybridisation: the Evidence of the Minoan Presence in the North East Aegean between the Middle and Late Bronze Age", in: N. C. Stampolidis, Ç. Maner and K. Kopanias (eds), *NOSTOI. Indigenous Culture, Migration and Integration in the Aegean Islands and Western Anatolia during the Late Bronze Age and Early Iron Age*, Koç Üniversitesi Yayınları 58, Koç University Press, Istanbul, 2015, 387-420.

L. Girella and P. Pavúk, "The nature of Minoan and Mycenaean involvement in the Northeastern Aegean", in: E. Gorogianni, P. Pavúk and L. Girella (eds), *Beyond Thalassocracies Understanding Processes of Minoanisation and Mycenaeanisation in the Aegean*, Oxford and Philadelphia, Oxbow Books, 2016, 15-42.

D.B. Grammenos, *Νεολιθική Μακεδονία*, Δημοσιεύματα του Αρχαιολογικού Δελτίου Αρ. 56, Υπουργείο Πολιτισμού, Αθήνα, 1997.

T.D. Hall, P.N. Kardulias and C. Chase-Dunn, "World-Systems Analysis and Archaeology: Continuing the Dialogue", *Journal of Archaeological Research*, 19, 3 (2010), 233-279.

I.S. Ivanov and M. Avramova, *Varna Necropolis: The Dawn of European Civilization*, Sofia: Agatho, 2000.

K. Kallintzi and S. Papadopoulos, "Ο αρχαιολογικός χώρος Διομήδειας Ν. Ξάνθης", *Το Αρχαιολογικό Έργο στη Μακεδονία και τη Θράκη* 21 (2007), 2010, 443-454

P.L. Kohl, "World-systems and modelling macro-historical processes in later prehistory: an examination of old and a search for new perspectives", in: T.C.

Wilkinson, S. Sherratt and J. Bennet (eds) *Interweaving Worlds. Systemic Interactions in Eurasia, c.7th to 1st Millennia BC*, Oxford, Oxbow Books, 2011, 77-86.

Ch. Koukouli-Chrysanthaki, "Οικισμός της Πρώιμης Εποχής του Χαλκού στη Σκάλα Σωτήρος Θάσου", *Το Αρχαιολογικό Έργο στη Μακεδονία και τη Θράκη* 1 (1987) 1988, 389-406.

Ch. Koukouli-Chrysanthaki, "Τα 'μέταλλα' της Θασιακής Περαίας", in: *Μνήμη Δ. Λαζαρίδη: Πόλις καί Χώρα στην Αρχαία Μακεδονία και Θράκη, Πρακτικά Αρχαιολογικού Συνεδρίου (Καβάλα 1986)*, Ελληνογαλλικές Έρευνες Ι. Θεσσαλονίκη, 1990, 493-514.

Ch. Koukouli-Chrysanthaki, *Πρωτοϊστορική Θάσος, Τα νεκροταφεία του οικισμού Καστρί*, Αθήνα: Δημοσιεύματα του Αρχαιολογικού Δελτίου, Αρ. 45, 1992.

Ch. Koukouli-Chrysanthaki and Y. Bassiakos "Non-slagging copper production of 5th millennium: The evidence from the Neolithic settlement of Promachon- Topolnica (Estern Macedonia, Greece)", in: *8th EAA Annual Meeting 24-29 September 2002, Thessaloniki, Abstracts book*, Thessaloniki, 2002, 193-194.

Ch. Koukouli-Chrysanthaki, H. Todorova, I. Aslanis, I. Vajsov and M. Valla, "Promachon-Topolnica: A Greek-Bulgarian archaeological project", in: H. Todorova, M. Stefanovich and G. Ivanov (eds), *Proceedings of the International Symposium "Strymon Praehistoricus", The Struma/Strymon River Valley in Prehistory*, Sofia, Gerda Henkel Stiftung, 2007, 43-78.

Ch. Koukouli-Chrysanthaki and S. Papadopoulos, "The island of Thasos from the Neolithic to the Early Bronze Age. Excavation data and absolute dates", in: Z. Tsirtsoni (ed.), *The Human Face of Radiocarbon, Reassessing Chronology in prehistoric Greece and Bulgaria, 5000-3000 cal. BC*, Travaux de la Maison de l'Orient et de la Méditerranée, 69, 2016, 339-358.

K. Kristiansen, *Europe Before History*. Cambridge: Cambridge University Press, 1998.

P. Leshtakov, "Early copper metallurgy on the west Black Sea coast: archaeological evidence on prehistoric exploitation of the Rosen ore field", in: *Where are the sites? Research, Protection and Management of Cultural Heritage*, Ahtopol, Bulgaria, Sozopol: Centre for Underwater Archaeology, Bulgaria, 2013, 35-54.

V. Leusch, S. ZÄuner, V. Slavčev, R. Krauss, B. Armbruster and E. Pernicka, "Rich metallurgists´(?) graves from the Varna I cemetery. Rediscussing the social role of the earliest metalworkers", in: A. Brysbaert and A. Gorgues (eds), *Artisans versus Nobility*. Leiden: Sidestone Press, 2017, 101-24.

D. Malamidou, Z. Tsirtsoni, P. Yiounni, L. Lespez, V. Kilikoglou and Tsolakidou, "Les poteries Néolithiques à décor peint "noir sur rouge" en Grèce du Nord: matières premières et production", *BCH* 130, 2006, 571-611.

D. Matsas, "Samothrace and the NE Aegean: the Minoan connection", *Studia Troica* 1, 1991, 159-179.

D. Matsas, "Minoan long-distance trade: a view from the NE Aegean", in: R. Laffineur and W.R. Niemeier (eds), *Politeia. Society and State in the Aegean Bronze Age, Aegeum* 12: II, 1995, 235-247.

D. Matsas, "Problems in Island Archaeology: Towards an Archaeology of Religion on Samothrace", in: O. Palagia and B.D. Wescoat (eds), *Samothracian Connections: Essays in Honour of James R. McCredie*, Oxford, Oxbow Books, 2010, 33-49.

D. Matsas, L. Girella and P. Pavúk, "Minoanization or Hybridization? Aspects of the Minoan Presence at Mikro Vouni (Samothrace) in the Middle Bronze Age", *Unpublished paper presented at XIth Diethnes Kritologhiko Synedrio*, October 2011, Rethymno, 2011, 21-27.

V. McGeehan-Liritzis, "The relationship between metalwork, copper sources and the evidence for settlement in the Greek Late Neolithic and Early Bronze Age", *Oxford Journal of Archaeology* 2, 1983, 147-180.

N. Nerantzis, Y. Bassiakos and S. Papadopoulos, "Copper metallurgy of the Early Bronze Age in Thasos, north Aegean", *Journal of Archaeological Science: Reports* 7, 2016, 574-580.

J. Palermo, *The impact of iron technology on the economy of the Aegean and Cyprus from 1200-850 BCE*, Unpublished PhD Thesis, University of Oxford, 2018.

S. Papadopoulos and D. Malamidou, "Προϊστορικός Οικισμός Λιμεναρίων: Το χρονικό της Έρευνας", in: S. Papadopoulos and D. Malamidou (eds), *Πρακτικά Ημερίδας Δέκα Χρόνια Ανασκαφική Έρευνα στον Προϊστορικό Οικισμό Λιμεναρίων Θάσου*, ΥΠΠΟΤ-ΙΗ' ΕΠΚΑ, Θεσσαλονίκη, 2012, 37-50.

Z.D. Papadopoulou, "Akrotiraki and Skali: New Evidence for EBA Lead/ Silver and Copper Production from Southern Siphnos", in: P.P. Betancourt and S.C. Ferrence (eds), *Metallurgy: Understanding How, Learning Why. Studies in Honor of James D. Muhly*, Philadelphia, 2011, 149-156.

E. Photos, Ch. Koukouli-Chrysanthaki, R. F. Tylecote and G. Gialoglou, "Precious Metals Extraction in Palaia Kavala, NE Greece, An Archaeometallurgical Attempt to Locate Skapte Hyle", in: *International Old World Archaeometallurgy Conference, Der Anschnitt*, Heidelberg: Max Plank Institut fur Kern Physik, 1989, 179-190.

M. Porčić, "Evaluating Social Complexity and Inequality in the Balkans between 6500 and 4200 BC", *Journal of Archaeological Research* 27, 2019, 335-390.

M. Radivojević, Th. Rehren, E. Pernicka, D. Sljivar, M. Brauns and D. Boric, "On the origins of extractive metallurgy: new evidence from Europe", *Journal of Archaeological Science* 37, 2010, 2775-2787.

M. Radivojević, B.W. Roberts, M. Marić, J. Kuzmanović Cvetković and Th. Rehren, *The Rise of Metallurgy in Eurasia. Evolution, Organisation and Consumption of Early Metal in the Balkans*, Archaeopress, Oxford, 2021.

M. Radivojević and B.w. Roberts, "Balkan metallurgy and society, 6200-3700 BC", in: M. Radivojević, B.W. Roberts, M. Maric, J. Kuzmanović Cvetkoviv and Th. Rehren (eds), *The Rise of Metallurgy in Eurasia. Organisation and Consumption of Early Metal in the Balkans*, Archaeopress, Oxford, 2021, 11-37.

Th. Rehren , P. Penkova and E. Pernicka, "The Chalcolithic Copper Smelting at Akladi Cheiri", in: R. Kraus, E. Pernicka, R. Kunze, K. Dimitrov and P. Leshtakov (eds), *Prehistoric Mining and Metallurgy at the Southeast Bulgarian Black Sea Coast*, Ressourcenkulturen 12, Tubingen University Press, 2020, 141-160.

C. Renfrew, M. Gimbutas and E.S. Elster, *Excavations at Sitagroi. A Prehistoric Village in Northeast Greece*, Volume 1, Los Angeles, California, 1986.

B.W. Roberts, C.P. Thornton and V.C. Pigott, "Development of metallurgy in Eurasia", *Antiquity* 83, 2009, 1012-1022.

N. Ryndina, G. Indenbaum and V. Kolosova, "Copper production from polymetallic sulphide ores in the northeastern Balkan Eneolithic culture", *Journal of Archaeological Science* 26, 1999, 1059-1068.

C.P. Thornton and B.w. Roberts, "Introduction: The beginnings of Metallurgy in Global Perspective", *Journal of World Prehistory*, 22, 2009, 181-184.

M. Tiverios, "Greek Colonisation of the northern Aegean", in: G. R. Tsetskhladze (ed.), *Greek Colonization, An Account of Greek Colonies and other Settlements Overseas*, Leiden-Boston, Brill, 2008, 1-154.

H. Todorova, "Die Anfänge der Metallurgie an der westlichen Schwarzmeerküste", in: A. Hauptmann, E. Pernicka, Th. Rehren and Ü. Yalçin (eds), *The Beginnings of Metallurgy. Proceedings of the International Conference, Bochum 1995, Bochum, Deutsches Bergbau-Museum, Der Anschnitt*, Beiheft 9, 1999, 237-246.

R. Treuil, "Dikili Tash: Village Prehistorique de Macedoine Orientale, I. Fouilles de Jean Deshayes (1961-1975)", *BCH, Supplement XXIV*, 1992, Athens.

M. Vavelidis, G. Christofidis and V. Melfos, "Ancient gold mines near Krinides in Philippi area (Macedonia, N. Greece)", *Archaeometrie und Tenkmalpflege*, Wien, 1997, 84-86.

G.A. Wagner and G. Weisgerber, *Antike Edel- und Buntmetallgewinnung auf Thasos, Der Anschnitt*, Beiheft 6, Bochum: Bergbau-Museum, 1988.

K. Zachos, Η μεταλλουργία στην Ελλάδα και στη ΝΑ Ευρώπη κατά την 5η και 4η χιλιετία, in: N. Papadimitriou and Z. Tsirtsoni (eds), *Η Ελλάδα στο ευρύτερο πολιτισμικό πλαισίο των Βαλκανίων κατά την 5η και 4η χιλιετια*, Μουσείο Κυκλαδικής τέχνης, Ίδρυμα Ν.Π. Γουλανδρή, 2010, 77-91.

A.G. Zannis, Le pays entre le Strymon et le Nestos: géographie et histoire (VIIe-IVe siècle avant J.-C.), *Meletemata* 71, Athènes, 2014.

2. DIFFERENT SCALES OF ARCHAEOMETALLURGICAL RESEARCH AT THE SITES OF DHASKALIO AND KAVOS ON KEROS: SITE SURVEY, EXCAVATIONS, AND ISLAND-WIDE SURVEY

Myrto Georgakopoulou

Introduction

The presently uninhabited island of Keros, situated south of Naxos in the central Cyclades, has captured the interest of archaeologists and the general public at least since the early 1960, unfortunately following extensive looting and the appearance on the antiquities market of numerous fragmented Early Cycladic figurines with a purported Keros provenance.[1] Subsequent salvage investigations and excavations identified the location of the looted deposit to the western end of the island, known as Kavos[2] (Fig. 1). Systematic study of some of the looted material, now in museums and private collections, and comparison with finds from sanctioned excavations have suggested that some of the former were indeed connected to the Kavos looted area, now known as the Kavos Special Deposit North (SDN).[3] Small-scale excavations were also undertaken at that time on the small islet of Dhaskalio approximately 90m off the coast of Kavos (Fig. 1), where Early Bronze Age (EBA) remains were also identified.[4] Although the islet now stands clearly separate from Keros, geoarchaeological investigations, taking into account neotectonic submerging trends in the area and sea-level rise, strongly suggest that during the third millennium BCE it was connected to Keros through a thin strip of land.[5] The peculiar nature of the finds and the destruction of the site from looting made the understanding of this site difficult, prompting in 1987 a systematic intensive survey on Kavos, carried out as a collaboration between the Universities of Cambridge, Athens, and Ioannina.[6]

These earlier investigations on Keros had left a series of unanswered questions and the nature of the Dhaskalio and Kavos site, particularly that of the Special Deposit, was strongly debated with interpretations varying between a nodal settlement on Dhaskalio and associated cemetery on Kavos,[7] a sanctuary involving redeposition of funerary materials originally deposited on cemeteries elsewhere,[8] and a sanctuary associated with deposition of selected material unrelated to funerary activity.[9] With these questions in mind, a series of systematic research projects were initiated in 2006 under the umbrella of the Cambridge Keros Project, including excavations on Keros and Dhaskalio islet between 2006-2008 (Director: C. Renfrew);[10] the Keros Island Survey in 2012-2013 (Directors: C. Renfrew, M. Marthari, and K. Dellaporta); and the Keros-Naxos Seaways project between 2015-2018, which involved a systematic survey on southeast Naxos and on Kato Kouphonisi island (Directors: C. Renfrew, M.J. Boyd, and D. Athanasoulis) and further excavations on Dhaskalio and Kavos (Directors: C. Renfrew and M.J. Boyd).[11] The site complex includes several different areas, whose names have slightly varied over the years; these areas are annotated with their currently finalized names, as used in this paper, in Fig. 1.

Study of the metal and metallurgical finds from Keros commenced in 2001 with the study of the material from the 1987 survey[12] and has been an

1 See for example Getz-Gentle 2008; Getz-Preziosi 1983; Sotirakopoulou 2005, 31-37; Renfrew 2015, 399-403 and references within.

2 Doumas 1964; 2013; Hadji-Vallianou 1975; Renfrew 1972, 531-2; Zapheiropoulou 1968a; 1968b; 1975; Zapheiropoulou *et al.* 2007.

3 Getz-Gentle 2008; Renfrew 2008; Sotirakopoulou 2005; 2008.

4 Doumas 1964; 2013.

5 Bassiakos and Doumas 1998; Dixons and Kinnaird 2013.

6 Broodbank 2000a; Renfrew *et al.* 2007.

7 Broodbank 2000a; 2000b, 223-236; Doumas 1964.

8 Bassiakos and Doumas 1998; Doumas 1990.

9 Renfrew 1984; 1991.

10 Renfrew *et al.* 2013; 2015; 2018a; Sotirakopoulou 2016.

11 Boyd *et al.* 2021; Renfrew *et al.* 2018b.

12 Georgakopoulou 2007a; 2007b.

Fig. 1. Perspective view of western Keros, showing archaeological zones mentioned in the text. Scale is approximate and relates to the islet of Dhaskalio. Position of Keros in the central-southern Aegean in the insert.

integral part of the field projects that followed.[13] This paper brings together the results of this ongoing archaeometallurgical study focusing on how our understanding of metallurgical practices on the site has developed with the field research projects of different scale and nature that have been undertaken. The focus here are the sites of Dhaskalio and Kavos and the island of Keros more generally, so the material from the surveys on SE Naxos and Kato Kouphonisi will not be considered.

Earliest excavations and the 1987 survey on Kavos

The earliest finds relating to metallurgy date to the very first explorations on Kavos and Dhaskalio by C. Doumas and include a fragment of a blowpipe tip or tuyère from Kavos[14] and a large fragment of raw, unworked copper metal from Dhaskalio.[15] The 1987 survey focused on Kavos with a less intensive collection taking place on a promontory to the north of the SDN in an area initially labelled as Kavos North and subsequently referred to as Kavos Promontory. The archaeometallurgical assemblage from the survey included a range of materials; the majority were poor iron minerals, but there were also copper slags, copper metal spills or shapeless fragments, a few metallurgical ceramics, and a single litharge fragment.[16] The analytical examination of this material yielded some unexpected results in the context of our understanding of EBA metallurgy in the Cyclades. The slags were found to be clearly copper smelting slags and not by-products from secondary metalworking, as would have been expected from a settlement context so far away from any known relevant ore sources.[17] The slags could be divided into two groups; the significantly more numerous Group 1 was associated with the production of pure copper, mostly seen as matte prills in the slags, while Group 2 was found to be associated with the production of arsenical copper rich in lead.

13 Georgakopoulou 2013; 2015; 2018; Renfrew *et al.* 2018b.

14 Georgakopoulou 2015, 447.

15 Doumas 2013, 85-6; Georgakopoulou 2013, 688.

16 Georgakopoulou 2007.

17 The main ore sources for the prehistoric Cyclades are considered to be located on the western Cycladic islands of Kythnos, Seriphos and Siphnos, as well as on SE Attica on the Greek mainland at the large polymetallic deposit of Lavrion (see discussion and references in Georgakopoulou 2016).

This proposal was also supported by the study of metallurgical ceramics, that were primarily found to be associated with smelting, with the exception of a single sherd (KK6), whose slagged layer composition was more consistent with melting of copper rich in arsenic and lead.[18] Examination of the single litharge fragment showed that it was not pure lead oxide but contained a significant content (c. 50%) of gangue oxides, so it was recognized as litharge-impregnated hearth lining or cupel material.[19] This opened the possibility that cupellation was also practiced on the site and pointed to a particular technology that is now known from numerous sites across the Aegean of the same period with examples from Attica,[20] the Cyclades,[21] and Thasos,[22] as well as earlier examples beyond the Aegean.[23]

Field study or a visit to Keros was not feasible at the time, but the systematic collection strategies of the project[24] permitted a consideration of the spatial distribution of metallurgical practices across the investigated area.[25] The first observation was that metallurgical remains were almost completely absent from the more intensively collected area of the SDN. Most of the finds from this area were poor iron mineral fragments, so it was suggested that these may have been part of the local bedrock and would have been more intensely broken down and churned over through the looting and subsequent sanctioned excavations in this part of the site. These were therefore considered more likely to be geological rather than archaeological material. The question remained whether this perceived absence of metal-related finds at the SDN was real or whether any metal present had also formed part of the looted assemblage. The second important point was the significantly higher concentration of metallurgical material in the Kavos Promontory area, despite it having been far less intensively surveyed. The proposal was thus made that a metal production site was more likely located here.

18 Georgakopoulou 2007, 392-3.

19 Georgakopoulou 2007, 394-5.

20 See Georgakopoulou *et al.* 2020 and references within.

21 e.g. Bassiakos *et al.* 2013; Papadopoulou 2011.

22 Bassiakos *et al.* 2019.

23 Hess *et al.* 1998; Pernicka *et al.* 1998; 2011.

24 Whitelaw 2007.

25 Georgakopoulou 2007, 398, Fig. 11.24.

Excavations on Kavos and Dhaskalio 2006-2008

The excavations undertaken at Kavos between 2006-2008 focused on an area suspected to host a second undisturbed deposit similar to the SDN, an expectation that was indeed verified in the area now known as Special Deposit South (SDS).[26] Hundreds of fragments of marble Cycladic figurines and bowls were recovered, together with large quantities of pottery, obsidian and other materials. This deposit and by extension SDN, are now understood to be part of an Early Cycladic maritime sanctuary, where deposition of parts of marble figurines and vessels, and possibly other materials, fragmented elsewhere, was taking place as part of ritual activities.[27] Interestingly, despite the rich material finds from the SDS, metal and metallurgical finds are extremely rare: three copper-based fragments and a single silver ring from topsoil, the latter perhaps modern, as well as eight small fragments recognized as metallurgical remains including primarily metallurgical ceramics and shapeless copper metal.[28] This picture, which stands in stark contrast with the wealth of relevant finds elsewhere on the site (see below), confirmed the picture from the 1987 survey that metal did not form part of the materials deposited during the ritual activities on the Kavos sanctuary. The suspicion that iron mineral fragments found in different parts of the surveyed Kavos area and in particularly high concentrations in the area of the SDN during the 1987 survey were part of the local geology was indeed confirmed through field observations. A number of poor iron mineralisations exist in the area of Kavos and fragments of these abound due to erosion of these occurrences.[29]

The area of the Kavos Promontory was re-visited and a more intensive survey, as well as limited excavation, was undertaken.[30] The distribution of metallurgical remains, primarily slags, very clearly increases towards the northern tip of the promontory, where a small smelting site was identified. The location and layout of the Kavos Promontory smelting site is very similar to other known smelting sites of this period in the Aegean, at the top of a slope, facing north, in the direction of the prevailing winds, obviously making use of these

26 Renfrew *et al.* 2015.

27 Renfrew 2013.

28 Georgakopoulou 2015, 329-331.

29 Brodie and Georgakopoulou 2015, 519, Fig. 19.20; Dixon and Kinnaird 2013.

30 Brodie and Georgakopoulou 2015.

Fig. 2. Four intact tuyères (blowpipe tips) from Dhaskalio (excavations 2006-2008).

to encourage burning.[31] An important question at this stage was to try and define the exact chronological phasing of these activities, particularly with reference to the chronology of Dhaskalio and the Kavos Special Deposits (see below). Significant erosion at the Kavos Promontory has resulted in shallow deposits and there was no clear stratigraphy in the trenches opened, while pottery from all three Phases (defined below) was recognized at the Promontory.[32] Thermoluminescence dating of metallurgical ceramics from the Kavos Promontory confirmed dating of the activities to the third millennium BCE, but the error margins were too large to allow more precise phasing.[33] Analyses of the slags and other metallurgical finds from Kavos Promontory confirmed the presence of the two previously identified slag groups, corresponding to production of pure copper and lead-rich arsenical copper. The much larger volumes of material now collected conclusively corroborated the predominance of the Group 1 slags.[34]

Extensive excavations were also undertaken on Dhaskalio, focusing primarily on the summit of the steep islet.[35] Three Phases of occupation were identified, with absolute radiocarbon dates suggesting the following ranges: Phase A (2750-2550BCE), Phase B (2550-2400BCE), and Phase C (2400-2240BCE).[36] Most excavated contexts were attributed to Phase C. Metal and metallurgical finds were recovered from all three phases, but the vast majority were attributed to Phase C, in line with the chronology of most excavated deposits. The metallurgical remains from Dhaskalio were of a different nature to those of the Kavos Promontory and consistent with secondary copperworking activities.[37] They included six tuyères (blowpipe tips),[38] whole (Fig. 2) and fragmentary, and approximately ten more metallurgical ceramic fragments, most undiagnostic in terms of shape, but at least two could be recognized as baking pan fragments (see below), two joining fragments as a possible crucible, and some were attributed to a ceramic mould. Numerous small slags and copper spills were also identified, as well as a single litharge fragment, the only find suggesting additionally lead-silver metallurgy. Analytical examination of samples taken from the copper-based materials confirmed that they were associated with working arsenical copper. Three larger smelting slags were also recovered from Dhaskalio, these were rounded and shaped to be used as stone tools (Fig. 3), clearly suggesting that they were at Dhaskalio for secondary usage, rather than as remains of smelting practised in-situ.[39] A single fragment of speiss, in this case ferrous arsenide and antimonide, was recovered from a Phase C context.[40] This was a particularly exciting find, especially given that a number of projects from Iran[41] to Turkey[42] and

31 e.g. Bassiakos and Philaniotou 2007; Betancourt 2006; Georgakopoulou *et al.* 2011.

32 Sotirakopoulou forthcoming.

33 Brodie and Georgakopoulou 2015, 518-520.

34 Georgakopoulou 2018, 503-515.

35 Renfrew *et al.* 2013.

36 Renfrew *et al.* 2012; Manning 2015.

37 Georgakopoulou 2013; 2018.

38 For a discussion and references on parallels of these across the Aegean see Georgakopoulou 2013, 671-2.

39 For other Aegean parallels see Georgakopoulou 2013, 672.

40 Georgakopoulou 2013, 684-5; 2018, 522.

41 Rehren *et al.* 2012; Thornton *et al.* 2009.

42 Mehofer 2014.

Fig. 3. Three rounded large smelting slags from Dhaskalio (excavations 2006-2008) believed to have been used as tools.

Crete[43] have brought forward evidence that from the late 4th millennium BCE speiss was deliberately produced and used as an alloying component in the production of arsenical copper. At Dhaskalio Kavos arsenical copper production has been documented, not least with the Group 2 slags at the Kavos Promontory, but the pathway for production of this alloy remains unclear. Different pathways to produce arsenical copper include smelting of mixed copper-arsenic ores, mixing of copper-rich and arsenic-rich ores during smelting, or alloying copper metal with arsenic-rich minerals. The analytical examination of the Group 2 slags and other materials from the Kavos Promontory had not provided any characteristic features to distinguish between these three possibilities. The identification of speiss on Dhaskalio thus raises the possibility that such material was mixed with copper metal for alloying purposes. There is, however, some discrepancy in the chemical composition of the speiss and the arsenic-rich slag that makes such an association less straightforward. The high antimony content of the single speiss fragment is not reflected in the Group 2 slags analysed so far[44], although variable quantities of antimony were indeed detected in some of the enclosed metal prills. The overall antimony content was, however, much lower than arsenic in the bulk of the slags, in contrast to the speiss fragment where the two elements are present in comparable percentages. This discrepancy does not preclude the possibility that speiss was indeed used to alloy copper and produce arsenical copper. The few Group 2 slags analysed from the Promontory were shown to have a very heterogeneous chemistry, being considered a single Group due to several shared characteristics, such as the presence of arsenical copper prills or the overall higher iron to silicon ratio compared to Group 1. In addition, only a single speiss fragment has been found on the site so far; it is possible that, if speiss was indeed used, such materials also had a range of compositions. Lastly, the behaviour of the two elements in such a high temperature system has not been considered in detail.

Metallurgical remains were recovered from nearly all excavated trenches dating to Phase C and no particular concentration could be identified at this stage to argue for the existence of specialized workshop areas.[45] The excavated contexts of the previous Phases were too confined for any spatial distribution analysis to be undertaken.

Beyond evidence for metalworking, the excavations on Dhaskalio also brought to light metal artefacts, primarily of copper (one from Phase A; three from Phase B; twenty-three from Phase C), but also of lead (five from Phase B and seventeen from Phase C), and gold (one each from Phases B and C).[46] Most copper-based artefacts are small and fragmentary, primarily fragments of pins, tweezers, or fish hooks, but the identification of three tools recovered together in a room in the large hall at the summit of Dhaskalio (Fig. 4) stood out. The small hoard consisted of a chisel, a single axe and an adze axe, the latter broken in two before deposition (Fig. 5). Examination of these artefacts showed that these were functional tools and showed some signs of weathering or even breakage in the case of the adze axe, suggesting these may have been gathered together awaiting repair. Elemental analyses of the copper-based artefacts from Dhaskalio concluded that arsenical copper was by far the dominant alloy used on the site.[47] Only the single axe from the hoard was found to be made of tin bronze. This picture is consistent with most other assemblages studied from Early Cycladic contexts,

43 Doonan *et al.* 2007.

44 Georgakopoulou 2007, 386-389; Georgakopoulou 2018, 505-508.

45 Georgakopoulou 2013, 691, Fig. 32.36.

46 Georgakopoulou 2013.

47 Georgakopoulou 2018.

4

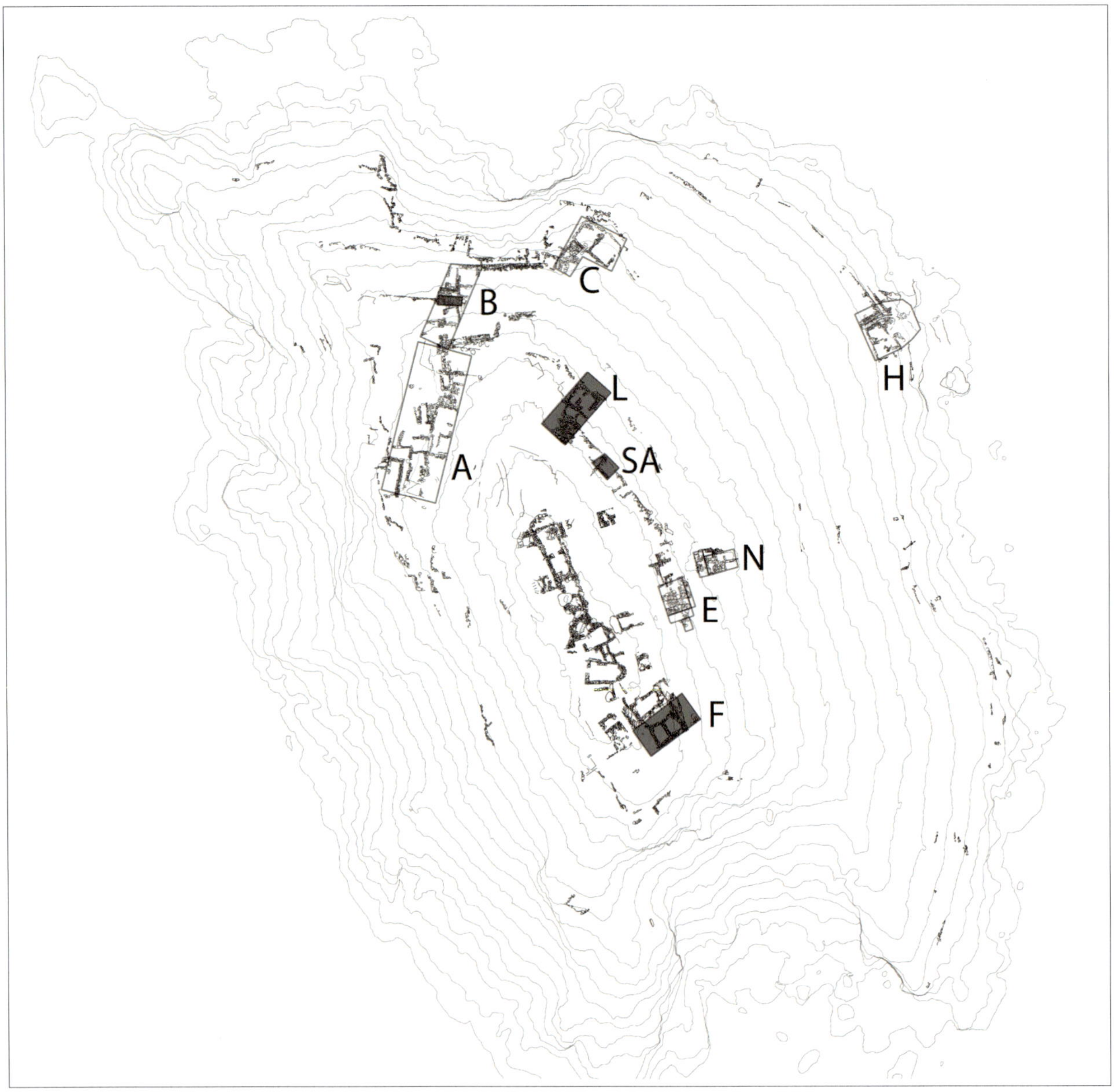

5

Fig. 4. Plan of Dhaskalio with trenches opened between 2016-2018 and structures revealed during the 2006-2008 excavations.

Fig. 5. Small hoard of copper-based tools found on the Dhaskalio summit (excavations 2006-2008).

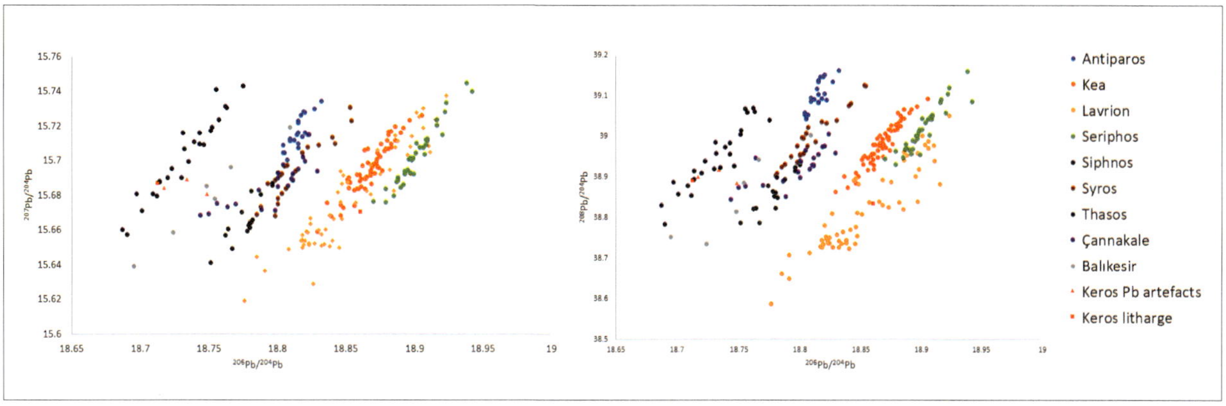

Fig. 6. Lead isotope diagrams for lead-based finds from Dhaskalio and Kavos with a selection of comparative fields (comparative data from https://oxalid.arch.ox.ac.uk/ and Begemann et al. 2003).

where elemental analyses have shown that arsenical copper was the most commonly used alloy with few tin bronze artefacts appearing in the region in the "Kastri" phase,[48] which is represented on Dhaskalio by Phase B.

The majority of lead artefacts were fragments of lead rivets for the repair of pottery. This is a common practice in the EBA Cyclades and numerous examples are known from many sites. A miniature lead shafthole axe was recovered from a Phase C context. Several EBA examples of miniature lead and copper tools are known from the Aegean and neighbouring regions.[49] Their small size and the softness of the metal in the case of lead makes it unlikely that these were functional tools. Many are from funerary contexts, particularly child burials, while at least the example from Dhaskalio and similarly the one from Palamari on Skyros have been found within settlements, where metalworking is known to have been practised. It has thus been proposed that these may have been given to children, who were associated with metalworkers, whether through family relations or apprenticeships. Lead is also a material often used to make weights and two examples from Dhaskalio, a spool and a cylindrical object with a depression on one side and 9 arranged dots on the other, may have held this function.[50]

The two gold artefacts, a flat, circular bead (Phase B) and a sheet fragment (Phase C), and further examples brought to light more recently (see below), are particularly significant in the context of Early Cycladic metallurgy, as gold is extremely rare on these islands during this period, often contrasted with the situation in the neighbouring northeastern Aegean and Crete.[51] All gold finds from Dhaskalio are very small and were mostly discovered during flotation residue sorting, highlighting the effect of rigorous recovery methods in the identification of small precious metal finds from settlements. Any attempt to compare the identification of gold artefacts on Dhaskalio in contrast to other large Cycladic settlements needs to take into account potential differences in field recovery methods.

In an attempt to address the question of raw material sources, both for the artefacts, but also for the in situ metallurgical activities undertaken at Dhaskalio and Kavos, a selection of artefacts and metallurgical remains were analysed for their lead isotope composition. A geological assessment of Keros concluded that the local geology could support the existence of small-scale copper mineralisations on the island.[52] Evidence for the existence of such occurrences was not, however, identified at the time and still is not known today (see also below). Comparison was therefore made with other known mineralisations in the broader region for lead (silver) and copper respectively.[53] Three lead artefacts and the two litharge fragments identified in the 1987 survey and the 2006-2008 excavations were analysed for their lead isotope compositions and are shown in Fig. 6 together with a selection of relevant sources from the broader region. The results are in general consistent

48 e.g. Gale and Stos-Gale 1989; Mangou and Ioannou 1997; Renfrew 1967.

49 See Georgakopoulou in press; Keskin in press; Marangou 1991 and references within.

50 Georgakopoulou 2013, 681-683.

51 Georgakopoulou 2013, 670.

52 Dixon and Kinnaird 2013.

53 For the assessment of the lead isotope data the databases of Aegean and Turkish ores from OXALID https://oxalid.arch.ox.ac.uk/default.html were used.

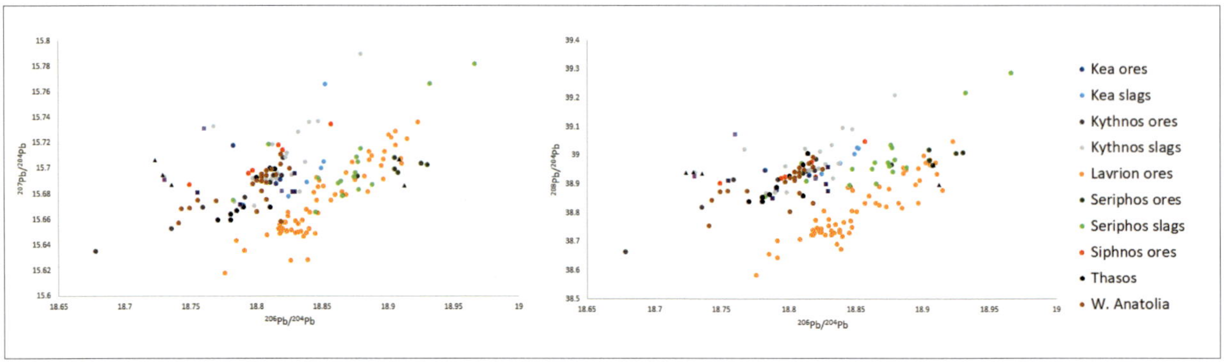

Fig. 7. Lead isotope diagrams for copper-based finds from Dhaskalio and Kavos with a selection of comparative fields (comparative data from https://oxalid.arch.ox.ac.uk/). Note that the tin bronze single axe is not depicted here.

with deposits known to have been exploited during the Early Bronze Age. Specifically, two of the artefacts and one of the litharge fragments are consistent with Siphnos, where lead-silver mining and production has been conclusively proven.[54] Interestingly the other litharge fragment has a distinctively different lead isotope composition, consistent with the Lavrion ore deposit, for which again there is evidence for EBA exploitation.[55] The third lead artefact has a lead isotope composition that may result from mixing lead from Siphnos and Lavrion, but is also somewhat consistent with lead ores from the region of Çannakale in Western Anatolia.

A range of different copper-based finds were analysed for their lead isotope composition: Group 1 and Group 2 slags from the Kavos Promontory, a slag tool and small slag from Dhaskalio, arsenical copper artefacts and the single tin bronze shafthole axe. The composition of the tin bronze axe is very distinct from the rest of the assemblage[56] and is comparable with some of the earliest tin bronzes found in the Aegean on sites such as Troy, Kastri on Syros[57] and Poliochni on Lemnos.[58] These are considered definite imports in the region from most likely distant but still presently unknown sources.[59] The remaining copper-related finds are broadly consistent with sources in the Aegean region (Fig. 7). Absolute provenancing in this case is complicated by significant overlap between the various sampled ore sources and slags. In the case of the western Cycladic islands, where EBA copper mining and production have been attested, slags sampled from large slag heaps on individual islands have been found to have a more widespread lead isotope composition, compared to ores from the same islands. It has been suggested that this distribution might indicate that ores from different islands within the western Cyclades were brought to these central smelting sites for processing.[60] This is indeed a very plausible scenario, however, the current lead isotope database for copper ores remains very limited for individual islands and the wider distribution of the slags may instead reflect wider isotopic fields for each island and possibly the co-existence of geologically distinct ore deposits within them.

Beyond absolute provenancing, there is some potentially interesting patterning within the data. Four out of the five Group 1 slags from the Kavos Promontory cluster close together and are broadly consistent with slags from the W. Cyclades (particularly Kythnos and Kea), but also with ores from Thasos and western Anatolia. The isotopic signature of the single Group 2 slag analysed is different. A single Group 1 slag, however, did plot between the Group 1 and 2 slags and may indicate a wider distribution of isotopic compositions. Still, based on the current data it is very interesting to note that although Group 1 slags were shown to be the majority on Kavos Promontory, none of the admittedly few Dhaskalio artefacts analysed here are directly consistent isotopically with them, while three arsenical copper artefacts were very close to the single Group 2 slag. Although the pattern is based on an overall small sample size, it does support the suggestion that the product of the Group 1 slags,

54 Papadopoulou 2011; Wagner and Weisgerber 1985.

55 Georgakopoulou *et al.* 2020 and references within.

56 It has been omitted from Fig. 7 to allow for better resolution of the remaining copper-based materials.

57 Stos-Gale *et al.* 1984.

58 Pernicka *et al.* 1990; 2003.

59 Stos-Gale 2016.

60 Gale and Stos-Gale 2008; Georgakopoulou 2016.

the unalloyed copper metal, was subsequently mixed with other materials (like speiss?), of a different lead isotope composition, to produce arsenical copper on Dhaskalio, supporting deliberate mixing of materials rather than use of mixed arsenic-copper minerals in the production of this alloy.[61] The other two copper-based artefacts have a very different isotopic composition, one being consistent with Lavrion and Seriphos ores and the other not really closely compatible with any of the sources considered here.

The slag tool from Dhaskalio has a lead isotope composition which is very compatible with the Group 1 slags. Its chemical composition was, however, found to be distinctly different with, among others, a much higher manganese content, lower calcium and enclosed arsenical copper prills. For this it seems more likely that the slag was brought to Dhaskalio from beyond Keros itself to be used as a tool. The other small slag from Dhaskalio has a chemical and isotopic composition that could be broadly consistent with the Promontory Group 2 slags, particularly given the large chemical variability seen amongst the few Group 2 slags analysed.

Keros Island Survey (2012-2013)

All the accessible areas on the island of Keros were systematically and intensively walked and representative archaeological finds of different types were collected during the survey undertaken between 2012-2013. Given its tendency to corrode, metal is not well preserved on the surface and, even when metal artefacts are found on surveys, they are usually fragmentary and not typologically diagnostic, making them impossible to date. On the contrary, metallurgical remains preserve well and are frequently recovered from surface surveys in the Aegean region.[62] Thus, in the context of the Keros archaeometallurgical investigation, the survey was of particular interest to address two main questions relevant to the metallurgical practices witnessed at Dhaskalio and Kavos. First, whether there were other sites in the proximity or further afield on Keros, where metallurgical activities were undertaken. Secondly, whether there may have been any small-scale mineralisations on the island that may have been the sources of at least some of the raw materials exploited in the metallurgical activities at Dhaskalio and Kavos (e.g. copper or arsenical minerals). Given the very limited scale of such resources on the Cyclades in general and the fact that any exploitation would have rendered them even less visible, a more systematic survey would be in a better position to identify any remaining traces and address the question of local resources conclusively.

The picture from the survey was clear. Although a number of locations with EBA material have been identified on Keros, metallurgical material similar to those known from Dhaskalio and Kavos, primarily slags and copper spills, were only found near the coast to the north of the Kavos Promontory, while a couple of copper slags were also found further to the east along the coast. This suggests that metallurgical activities were taking place on other locations to the north of the Kavos Promontory, but given the proximity these are clearly associated with the Dhaskalio and Kavos site complex. A few iron slags were collected further to the east in the most fertile part of the island around the areas of Gerani and Konakia, where material finds from various prehistoric and historic periods demonstrate multi-period habitation or more broadly activity. The iron slags are clearly associated with later historic periods. Similar material had been recognized previously and a small number of slags had been analysed.[63] On the basis of this distribution, it is clear that metallurgical activities were only undertaken in the broader area of the Dhaskalio and Kavos sites during the Early Bronze Age.

With regards to the second question, copper, lead or other relevant minerals were not found amongst the collected material. The field teams did, however, identify several small-scale open emery quarries in east central Keros, of as yet unclear date, strongly suggesting that if prehistoric mining activities had been undertaken on the island, their remnants would have been recognized in the field. The survey data thus confirmed that ores relevant to the metallurgical activities undertaken on Dhaskalio and Kavos are not available on Keros and would have been imported by maritime routes.

Excavations on Dhaskalio and Keros (2016-2018)

In the most recent excavation campaigns, the focus has been primarily on Dhaskalio, but some limited excavation was also undertaken on Keros, to the north of the Kavos Promontory. The finds from these investigations, metallurgical as well as all else, are still under study, so the results presented here are of a preliminary nature. The metal and metallurgical finds have been studied macroscopically and representative

61 For a fuller discussion see Georgakopoulou 2018, 530.

62 Georgakopoulou 2014.

63 Bassiakos and Doumas 1998.

Fig. 8. Baking pan rim sherd with evidence for metallurgy (a) external surface (b) internal surface.

samples have been selected for analyses, which are now underway.

The most recent excavations on Dhaskalio have added significantly to our understanding of this impressive site. Contrary to the 2006-2008 campaign, which focused on the summit of Dhaskalio, revealing contexts dating primarily to Phase C, trenches were now spread in different parts of the islet below the summit, where it became clear that the main Phases of activity on Dhaskalio were earlier, mainly of Phase B, as well as Phase A. Impressive architectural remains reveal a well-planned layout including terracing, drainage, and staircases, as well as buildings with walls constructed mainly of imported stones.[64]

The finds relating to metallurgy, both in terms of artefacts and metalworking remains, are impressive, truly surpassing in number and variety anything known from settlement contexts of the same period in the Cyclades. In terms of metal artefacts, the collection includes copper-based (c. 155), lead (c. 125), gold (7), and silver (3) finds, the vast majority being small fragments or small items like pins, needles, punches, lead clamps and beads or wire for the precious metals. A second miniature lead axe was recovered, this time a shaft hole single axe. Many of the artefacts, including the majority of the small precious metal finds, were recovered from sorting flotation residue, highlighting once again the importance of rigorous recovery methods for their identification during excavation.

The collection of metallurgical remains undeniably reveals Dhaskalio as a major metalworking centre for the EBA Aegean. The largest broad category of finds are the metallurgical ceramics, that is ceramics that are associated with metallurgy either due to their type or due to the identification of clear metallurgical residue on their surface. Amongst them the most common find are sherds of a vessel known as baking pan or hearth, a ceramic type recognized previously in substantial quantities from all three Phases on Dhaskalio and with parallels from various contexts in the EBA Cyclades.[65] These are large open vessels of various sub-types, with thin bases and usually flaring walls with rim diameters around 50-55cm. They are thought to be portable hearths for domestic activities relating primarily to cooking. On Dhaskalio specifically, however, some at least of these vessels were clearly associated with metallurgy, as attested by signs of intense burning and green copper mineral residues (Fig. 8). A few examples had been identified already during the 2006-2008 excavations and a single sherd had been analysed.[66] Higher calcium, magnesium and copper levels were consistent with absorption of fuel ash and copper processing. Exactly how these vessels were used in the metalworking activities on Dhaskalio remains to be investigated through the study of the many more examples now recovered. Doumas[67] has previously proposed that the toolkit of a metalsmith in the EBA Cyclades may have consisted of a complex assemblage of known Early Cycladic vessels. In this proposed reconstruction the baking pans would act as heating trays on which crucibles containing the metal to be molten would be placed and presumably heated internally and externally. Such a function certainly seems very plausible for these large vessels, although it should be noted that the particular combination of

64 Boyd *et al.* forthcoming.

65 Sotirakopoulou 2016, 36-41, 97-103, 221-229.

66 Georgakopoulou 2018; 520.

67 Doumas 2011.

vessels as proposed by Doumas is not really known from a single context with evidence for metallurgy. The ongoing study of the extensive ceramic and metallurgical ceramic assemblages of Dhaskalio will shed more light on this matter. Discriminating between baking pan sherds associated with metallurgy and those which are not is not always straightforward. Signs of burning and green copper-based residues are not consistent throughout the surfaces of the larger sherds, making it clear that some sherds from vessels that had been involved in metallurgical processes may not bear telling macroscopic features. In an attempt better to discriminate those associated with metallurgy, a comprehensive analysis of the entire assemblage of baking pan fragments using a portable x-ray fluorescence spectrometer (pXRF) is planned in the near future. To give a preliminary sense of the scale of the assemblage, at this stage more than 350 sherds are attributed to metallurgical processes.

Contrary to the abundance of baking pan fragments, the number of sherds identified as potential crucible fragments is small (c. 20). Admittedly, absolute numbers are not very indicative at this stage, as the size of the sherds varies and the baking pans are very large vessels that would generate many more fragments upon breaking. Until a minimum vessel number can be estimated, direct comparison is not possible, but still the overall impression is that the baking pans associated with metallurgy presently far exceed possible crucibles. All of the sherds identified as crucibles are clearly associated with metallurgy, as attested by intense burning or clear metallurgical residue. They usually show curvature indicating a rounded vessel and are thus not fragments of baking pans. All of the crucible sherds are small and it is difficult to establish their type.

Another common metallurgical ceramic find are blowpipe tips or tuyères, similar to those identified previously on Dhaskalio (see above). Approximately thirty-five examples, fragments or intact, are known from Dhaskalio; together with seven examples from previous excavations this constitutes one of the largest, if not the largest, assemblage of such tools known from a single site to date.

Beyond ceramics, a range of other materials testify to local metalworking activities. Several slag fragments were found (c. 75), the vast majority are small, below ten grams, with few larger exceptions. At least two large, rounded slags were identified; their characteristics suggest that these were on Dhaskalio for a secondary function as tools, similar to examples discussed previously. Three stone moulds, two broken and one intact, are the first of their kind recovered from Dhaskalio. The two broken ones are parts of bivalve moulds for midrib daggers and the other is a double-sided mould for casting a spearhead and a rectangular rod. All three have at least one smoothened edge, showing they were probably also used as stone tools, in the case of the broken ones after their breakage, as it is the broken surface that is smoothened. Minute, shapeless copper fragments abound on Dhaskalio; these are classified as copper spills, although some may be very weathered fragments of small artefacts. A new type of find in this campaign were small shapeless fragments of lead, classified as lead spills. Unlike copper spills, which are found throughout the excavated contexts, these have a very distinct distribution: almost all (c. 98%) were recovered from a few contexts in Trench A to the northwest of the summit. A few more litharge fragments from different contexts on Dhaskalio add to the evidence suggesting local practice of lead and silver metallurgy.

Metallurgical remains were unearthed from all excavated trenches in variable amounts. A comprehensive discussion of spatial distribution of the metallurgical finds and thus of the associated activities will need to await the completion of the study and dating of all relevant contexts. Still, a preliminary consideration is offered here, highlighting some of the most distinctive contexts. Three trenches in particular stood out. The first, briefly mentioned previously, is Trench A, which at $216m^2$ is the most extensive trench on Dhaskalio. One area revealed a high concentration of shapeless lead fragments, which have been macroscopically classified as lead spills. These finds strongly suggest the presence of lead working in this part of the site. Given the extensive evidence for the use of lead clamps to repair pottery on the site, it is possible that this was at least one of the lead-related activities undertaken here.

The vast majority of baking pan sherds so far identified as metallurgical (c. 95% of the total number) were recovered from a few contexts in Trench L, situated immediately to the northeast of the summit, along with approximately half of the crucible fragments and slags, and numerous copper spills, blowpipe tips and other related materials. The concentration of metallurgical remains from Trench L is unparalleled in any other excavated context on the islet, but the exact nature of these deposits remains to be assessed through full consideration of the excavated contexts and the other finds recovered.

A very different picture was noted in Trench H, situated at the lowest point excavated so far, to the east of the islet and near the most likely entry point to Dhaskalio from the narrow strip of land connecting it with Keros, as testified also from the identification of a well-made staircase in the northern part of the trench. Two hearths dug into the bedrock and part of a third one were uncovered here. Their association with metallurgy is clear, as specks of green copper minerals were identified on the walls of at least one of them, while all three showed elevated copper and lead readings based on portable X-ray fluorescence (pXRF) analysis undertaken during excavation.[68] The three hearths constitute unequivocal evidence for the existence of an in-situ metalworking workshop in Trench H. Despite these clear field features, the actual number of metallurgical remains from Trench H and the workshop contexts in particular are comparatively low, especially when contrasted with Trench L. Baking pan sherds with a clear metallurgical association are almost absent from Trench H and only a couple of crucible fragments have been identified, although it should be noted that a small number of initially undiagnostic sherds remain to be assessed further. Similarly, the number of slags from H is low, representing less than 15% of the total number retrieved on the site. In contrast, the number of copper spills recovered from Trench H is the highest amongst all excavated trenches and represents approximately 40% of the total number. Several other metallurgical finds were recovered from Trench H including the double-sided stone mould, some blowpipe tips, a lead ingot, a few litharge fragments, and several stone tools, which may have been used in metalworking activities.[69] This contrast between Trenches H and L is striking and very interesting. Further study of the metallurgical finds and other materials aims to address whether these differences may be due to a change in technology, the different nature of these areas, or other parameters.

Discussion

Metallurgy was clearly an important activity at the Dhaskalio and Kavos site complex, although interestingly metal does not appear to have formed part of the ritual depositions at the two Special Deposits on Kavos. There is also no doubt that the site was an important metallurgical centre for the region. The small copper smelting site on Kavos Promontory remains the only primary metal production site known in the EBA Cyclades outside the western Cyclades, where the majority of ore sources exploited in this period are located.[70] Slags, apparently of an earlier Final Neolithic date, have also been recently reported from the island of Antiparos;[71] results from their analytical examination are awaited in order to clarify their nature. Secondary metalworking activities are attested in most major Early Cycladic settlements, but the scale recognized on Dhaskalio is currently unparalleled. Earlier excavations in the region employed significantly less rigorous recovery methods and were more selective in choosing which finds were kept; these approaches have certainly limited the nature and number of metallurgical finds reported and available for study.[72] Still, the volume of material unearthed through the recent excavation programmes on Dhaskalio is exceptional and the lack of similarly extensive assemblages on other major contemporaneous sites cannot be merely attributed to different field methodologies.

The absence of locally available ores suggests that these raw materials were brought to the island, most likely from the western Cyclades through maritime routes, a practice that also finds parallels at the larger EBA smelting site of Chrysokamino on Crete.[73] The amount of copper resulting from these activities would be comparatively small; the slag heaps on some of the western Cycladic islands are substantially larger.[74] The importance of these activities probably lay in the symbolism and status they carried in the voyage associated with bringing the raw materials and the knowledge and skill involved in the smelting.

In terms of the mobility involved in the metallurgical practices at Dhaskalio and Kavos, it is not only copper ores that found their way to the site. A number of other metallurgical raw materials or intermediary by-products are also clearly imported. They include speiss and any other raw material possibly involved in arsenical copper production, whether arsenical or arsenical copper minerals. Litharge is clearly imported also, in fact lead isotope analyses of two fragments suggested the material arrived from both major southern Aegean sources of the period, Lavrion and Siphnos. It is not known if litharge arrived on Keros as argentiferous lead and underwent cupellation in situ leaving behind litharge, or as litharge for other

68 These analyses were undertaken by Matt Lester and Roger Doonan, who ran the soil chemistry analyses across the site.

69 The tools are currently being studied by Diego Capra and Giulio Lucarini.

70 Georgakopoulou 2016.

71 Papadopoulou 2017a, 115; 2017b, 355.

72 Georgakopoulou 2007b.

73 Betancourt 2006.

74 Georgakopoulou 2016.

secondary usage.[75] To these materials we also need to consider clays or ready-made metallurgical ceramics arriving to Keros, as preliminary data from these ceramics show that none of them are local, as with the pottery on the site,[76] as well as stone or slag tools, also imported. Lastly, the question of fuel availability for these activities forms part of the current study and it is possible that this is another component that needed to be imported. Metal was also brought to the site, as local production certainly was not of a large scale. This would include raw metal to be formed into artefacts, but it is also possible that finished products arrived as well. In short, consideration of all stages of metal production on Dhaskalio and Kavos testify to extensive mobility of raw materials, intermediate and final products from a range of sources, emphasising the site's extensive connectivity and nodal position, also evident in other categories of finds.

In closing we return to the primary aim of this overview, which was to illustrate how studying the archaeometallurgical assemblages recovered from different types and scales of field project has been shaping our understanding of metallurgical practices at Dhaskalio and Kavos. In this context the wealth of information revealed already from the in-depth study of the humble collection of the 1987 surface survey on the Kavos area is striking. The study of these finds, which fitted in a medium-sized Tupperware container (contrasting with crates of material from more recent expeditions), revealed a diversity of activities (including surprising evidence for primary copper production), drew attention to Kavos Promontory as a potential smelting location, and proposed that metal most likely did not form part of the activities associated with the Special Deposit. Subsequent excavations on Kavos and Dhaskalio not only confirmed and enhanced these preliminary conclusions, but additionally brought forward evidence for working and usage of metal on Dhaskalio, particularly from the later Phase C. The more extensive recent excavations on the islet expanded the number and range of material finds, which now came from a wider range of contexts both in terms of chronology, spanning all Phases of occupation, and in terms of spatial distribution across Dhaskalio. The study of these finds will now permit a clearer resolution of metalworking technology identifying potential chronological changes, as well as a more detailed consideration of the spatial organization of these activities across Dhaskalio in the different Phases of occupation. Lastly, the archaeometallurgical research also benefitted from the island-wide survey, which localized prehistoric metallurgical activities in the vicinity of Dhaskalio and Kavos only and confirmed the absence of local raw material sources on Keros. The conclusions drawn so far offer significant insight into the importance of this craft on the site itself, as well as the role of Dhaskalio and Kavos in metal production, working, and circulation more broadly in the region. The wealth of data emerging highlights the importance of detailed macroscopic and laboratory studies of entire assemblages of finds from different scales of excavations and surface surveys. Specialist studies during or immediately following fieldwork, where possible, can furthermore provide opportunities to maximize information gathering and can have a positive informative impact on the research focus of an ongoing or future field project.

Acknowledgments

I thank N. Nerantzis for the invitation to participate in this symposium and volume. I am indebted to the directors of the Cambridge Keros Project Prof. C. Renfrew and Dr. M. Boyd for the invitation to work on the project and their continuous support in many ways over the years. I am particularly grateful to Dr M. Boyd for his helpful remarks on an earlier version of this paper. Prof. T. Whitelaw first got me involved in the study of the Keros metallurgy and Profs. Th. Rehren and C. Broodbank supervised the early work as part of PhD research. The Cambridge Keros Project is a collaborative effort and the archaeometallurgical study has benefited greatly from interaction and collaboration with all participants and specialists. Permit requests for analysis have been granted by the Ephorate of Antiquities of Cyclades and the Hellenic Ministry of Culture through the administrative assistance of T. Gerousi at the British School at Athens. Analytical work was undertaken at the Institute of Archaeology at UCL, the Fitch Laboratory, UCL Qatar, and STARC at the Cyprus Institute. Lead isotope analyses were carried out by Prof. R. Frei and his team at the Institute of Geography and Geology of the University of Copenhagen. Funding for the analysis is gratefully acknowledged from the Institute of Aegean Prehistory (INSTAP), UCL Qatar, The Cyprus Research & Innovation Foundation under the project "The antecedents of urbanism in Europe (URBAN)", PI: E. Margaritis (The Cyprus Institute), and STARC at the Cyprus Institute.

75 See discussion in Georgakopoulou 2007, 394-5.

76 Hilditch 2013; 2018.

Bibliography

Y. Bassiakos and C. Doumas, "The island of Keros and its enigmatic role in the Aegean E.B.A.: A geoarchaeological approach", in: *Argyritis Gi*, Athens, 1998, 55-64.

Y. Bassiakos and O. Philaniotou, "Early copper production on Kythnos: archaeological evidence and analytical approaches to the reconstruction of the metallurgical process", in: P.M. Day and R.C.P. Doonan (eds), *Metallurgy in the Early Bronze Age Aegean*, Oxford, 2007, 19-56 [*Sheffield Studies in Aegean Archaeology* 7].

Y. Bassiakos, M. Georgakopoulou, and G.A. Wagner, "Οι νέες αρχαιομεταλλουργικές μελέτες στη Σίφνο", in: *Proceedings of the 4th Siphnian Symposium*, Athens, 2013, 45-64.

Y. Bassiakos, N. Nerantzis, and S. Papadopoulos, "Late Neolithic/Early Bronze Age metallurgical practices at Limenaria, Thasos: evidence for silver and copper production", *Archaeological and Anthropological Sciences* 11, 11 (2019), 2743-2757.

F. Begemann, S. Schmitt-Strecker, and E. Pernicka. "On the compositions and provenance of metal finds from Besiktepe (Troia)", in: G.A. Wagner, E. Pernicka, and H-P. Uerpmann (eds), *Troia and the Troad*, Berlin, 2003, 173-201.

P.P. Betancourt, *The Chrysokamino Metallurgy Workshop and its Territory*, Princeton, 2006 [*Hesperia Supplement* 36].

M.j. Boyd, R. Campbell, R.c.p. Doonan, C. Douglas, G. Gavalas, M. Gkouma, C. Halley, B. Hartzler, J.A. Herbst, H.R. Indgjerd, A. Krijnen, I. Legaki, E. Margaritis, N. Meyer, I. Moutafi, N. Pirée Iliou, D.A. Wylie, and C. Renfrew, "Open Area, Open Data: Advances in Reflexive Archaeological Practice", *Journal of Field Archaeology* 46, 2 (2021), 62-80.

M.J. Boyd, C. Renfrew, I. Legaki, E. Margaritis, G. Gavalas, and I. Moutafi (eds), *The Sanctuary on Keros and the Origins of Aegean Ritual Practice,* Vol. IX: *The antecedents of urbanism: excavations 2016-2018. The excavation on Dhaskalio and the architecture*, Cambridge, forthcoming.

N. Brodie and M. Georgakopoulou, "Investigations on Kavos Promontory", in: C. Renfrew, O. Philaniotou, N. Brodie, G. Gavalas, and M.J. Boyd (eds), *The Sanctuary on Keros and the Origins of Aegean Ritual Practice, Vol. II. Kavos and the Special Deposits*, Cambridge, 2015, 507-523 [*McDonald Institute Monographs*].

C. Broodbank, "Perspectives on an Early Bronze Age island centre: an analysis of pottery from Daskaleio-Kavos (Keros) in the Cyclades", *Oxford Journal of Archaeology* 19, 4 (2000a), 323-342.

C. Broodbank, *An Island Archaeology of the Early Cyclades*, Cambridge, 2000b.

J. Dixon and T. Kinnaird, "The geology of Kavos and Dhaskalio", in: C. Renfrew, O. Philaniotou, N. Brodie, G. Gavalas, and M.J. Boyd (eds), *The Sanctuary on Keros and the Origins of Aegean Ritual Practice, Vol. I. The Settlement at Dhaskalio*, Cambridge, 2013, 25-44 [*McDonald Institute Monographs*].

R.C.P. Doonan, P.m. Day, N. Dimopoulou-Rethemiotaki, "Lame excuses for emerging complexity in Early Bronze Age Crete: the metallurgical finds from Poros Katsambas and their context", in: P.M. Day and R.C.P. Doonan (eds), *Metallurgy in the Early Bronze Age Aegean*, Oxford, 2007, 98-122 [*Sheffield Studies in Aegean Archaeology* 7].

C. Doumas, "ρχαιότητες και μνημεία των Κυκλάδων", *Archaiologikon Deltion (Chronika)* 19B, (1964), 409-12.

C. Doumas, "Death", in: L. Marangou (ed.), *Cycladic Culture: Naxos in the 3rd millennium BC*, 93-5. Athens, 1990, 93-5.

C. Doumas, "Searching for the Early Bronze Age Aegean metallurgist's toolkit", in: P.P. Betancourt and S.C. Ferrence (eds), *Metallurgy: Understanding How, Learning Why. Studies in Honour of James D. Muhly, Philadelphia,* 2011, 165-179.

C. Doumas, "The Dhaskalio excavations of 1963", in: C. Renfrew, O. Philaniotou, N. Brodie, G. Gavalas, and M.J. Boyd (eds), *The Sanctuary on Keros and the Origins of Aegean Ritual Practice, Vol. I. The Settlement at Dhaskalio*, Cambridge, 2013, 79-86 [*McDonald Institute Monographs*].

N.H. Gale and Z.A. Stos-Gale, "Some aspects of early Cycladic copper metallurgy", in: C. Domergue (ed.), *Mineria y Metallurgia en las Antiguas Civilizaciones Mediterraneas y Europeas*, Madrid, 1989, 21-37.

N.H. Gale and Z.A. Stos-Gale, "Changing patterns in prehistoric Cycladic metallurgy", in: N. Brodie, J. Doole, G. Gavalas, and C. Renfrew (eds), *Horizon: A Colloquium on the Prehistory of the Cyclades*, Cambridge, 2008, 387-408.

M. Georgakopoulou, "The metallurgical remains", in: C. Renfrew, C. Doumas, L. Marangou, and G. Gavalas (eds), *Keros, Dhaskalio Kavos: The Investigations of 1987-88*, Cambridge, 2007a, 380-401 [*McDonald Institute Monographs*].

M. Georgakopoulou, "Metallurgical activities within Early Cycladic settlements: the case of Daskalio-Kavos", in: P.M. Day and R.C.P. Doonan (eds), *Metallurgy in the Early Bronze Age Aegean*, Oxford, 2007b, 123-134 [*Sheffield Studies in Aegean Archaeology* 7].

M. Georgakopoulou, "Metal artefacts and metallurgy", in: C. Renfrew, O. Philaniotou, N. Brodie, G. Gavalas, and M.J. Boyd (eds), *The Sanctuary on Keros and the Origins of Aegean Ritual Practice, Vol. I. The Settlement at Dhaskalio*, Cambridge, 2013, 667-692 [*McDonald Institute Monographs*].

M. Georgakopoulou, "Metallurgical remains from regional surveys of "non-industrial" landscapes: The case of the Kythera Island Project", *Journal of Field Archaeology* 39, 1 (2014), 67-83.

M. Georgakopoulou, "Metals and metallurgy (contributions in Chapters 10, 14, 15, 17)", in: C. Renfrew, O. Philaniotou, N. Brodie, G. Gavalas, and M.J. Boyd (eds), *The Sanctuary on Keros and the Origins of Aegean Ritual Practice, Vol. II. Kavos and the Special Deposits*, Cambridge, 2015, 329-331; 408; 446-447; 482 [*McDonald Institute Monographs*].

M. Georgakopoulou, "Metal production, working, and consumption across the sites at Dhaskalio and Kavos", in: C. Renfrew, O. Philaniotou, N. Brodie, G. Gavalas, and M.J. Boyd (eds), *The Sanctuary on Keros and the Origins of Aegean Ritual Practice: the excavations of 2006-2008, Vol. III. The Marble Finds from Kavos and Archaeology of Ritual*, Cambridge, 2018, 501-532 [*McDonald Institute Monographs*].

M. Georgakopoulou, Y. Bassiakos, and O. Philaniotou, "Seriphos surfaces: a study of copper slag heaps and copper sources in the context of Early Bronze Age Aegean metal production", *Archaeometry* 53, 1 (2011), 123-145.

M. Georgakopoulou, K. Douni, M. Ginalas, O. Kakavogianni, And Y. Bassiakos, "Recent finds from Final Neolithic and Early Bronze Age silver production in southeastern Attica", in: N. Papadimitriou, J. Wright, S. Fachard, N. Polychronakou-Sgouritsa, E. Andrikou (eds), *Athens and Attica in Prehistory. Proceedings of the International Conference, 27-31 May 2015*, Oxford, 2020, 185-92.

M. Georgakopoulou, S. Bonatsos, Z.a. Stos-Gale, N. Dimitriou, and Y. Bassiakos, "In the middle of it all: Early and Middle Bronze Age metallurgy at Palamari (Skyros) at the crossroads of northeast and southern Aegean traditions", in: Y. Bassiakos (ed.), *Prehistoric Metal Production in the Aegean- Material Evidence and Analysis*, Heidelberg, in press.

P. Getz-Gentle, "The Keros Hoard Revisited", *American Journal of Archaeology* 112, 2 (2008), 299-305.

P. Getz-Preziosi, "The "KerosHoard': introduction to an Early Cycladic Enigma", in: D. Metzler and B. Otto (eds), *Antidoron: Festschrift für Jürgen Thimme*. Karlsruhe, 1983, 37-44.

D. Hadji-Vallianou, "Κέρος", *Archaiologikon Deltion (Chronika)*, 30B, (1975), 327.

K. Hess, A. Hauptmann, H. Wright, and R. Whallon, "Evidence of Fourth Millennium BC Silver Production at Fatmalı-Kalecik, East Anatolia", in: Th. Rehren, A. Hauptmann, J. Muhly (eds), *Metallurgica Antiqua*, Bochum, 1998, 57-67 [*Der Anschnitt* 8].

J. Hilditch, "The fabrics of the ceramics at Dhaskalio", in: C. Renfrew, O. Philaniotou, N. Brodie, G. Gavalas, and M.J. Boyd (eds), *The Sanctuary on Keros and the Origins of Aegean Ritual Practice, Vol. I. The Settlement at Dhaskalio*, Cambridge, 2013, 465-82 [*McDonald Institute Monographs*].

J. Hilditch, "The fabric study of the pottery of Dhaskalio and Kavos", in: C. Renfrew, O. Philaniotou, N. Brodie, G. Gavalas, and M.J. Boyd (eds), *The Sanctuary on Keros and the Origins of Aegean Ritual Practice: the excavations of 2006-2008, Vol. III. The Marble Finds from Kavos and Archaeology of Ritual*, Cambridge, 2018, 445-493 [*McDonald Institute Monographs*].

L. Keskin, "Early Bronze Age metalworking in the İzmir region: new evidence for chronology and interaction", in: C. Doumas, A. Giannikouri, and O. Kouka (eds), *The Aegean Early Bronze Age: New Evidence, Proceedings of an International Conference held in Athens, April 11-14th* 2008, Athens, in press.

H. Mangou and P.V. Ioannou, "On the chemical composition of prehistoric Greek copper-based artefacts from the Aegean region", *The Annual of the British School at Athens* 92, (1997), 60-72.

S. Manning, "Radiocarbon dating and archaeology. History, progress and present status", in: R. Chapman, A. Wylie (eds), *Material Evidence: Learning from Archaeological Practice*, Abingdon, 2015, 128-158.

C. Marangou, "Social differentiation in the Early Bronze Age miniature metal tools and child burials", *Journal of Mediterranean Studies* 1, 2 (1991), 211-225.

M. Mehofer, «Metallurgy during the Chalcolithic and the Beginning of the Early Bronze Age in Western

Anatolia», in: B. Horejs and M. Mehofer (eds), *Western Anatolia before Troy. Proto-Urbanisation in the 4th Millennium BC? Proceedings of the International Symposium held at the Kunsthistorisces Museum Wien Vienna, Austria, 21-24 November, 2012*, Vienna, 2014, 463-490.

Z.D. Papadopoulou, «Akrotiraki and Skali: New Evidence for EBA Lead/ Silver and Copper Production from Southern Siphnos», in: P.P. Betancourt and S.C. Ferrence (eds), *Metallurgy: Understanding How, Learning Why. Studies in Honour of James D. Muhly, Philadelphia*, 2011, 149-156.

Z.D. Papadopoulou, «Sculptures from Akrotiraki, Siphnos and its Cemetery», in: M. Marthari, C. Renfrew, and M.J. Boyd (eds), *Early Cycladic Sculpture in Context*, Oxford, 2017a, 107-118.

Z.D. Papadopoulou, «Πρόσφατες αρχαιολογικές έρευνες στην Αντίπαρο», in: P. Triantafylidis (ed.), *Archaeological Projects in the Aegean Islands, International Conference, Rhodes 27 November-1 December 2013*, Vol. B, Mytilene, 2017b, 355-368.

E. Pernicka, F. Begemann, S. Schmitt-Strecker, and A. Grimanis, On the composition and provenance of metal artefacts from Poliochni on Lemnos, *Oxford Journal of Archaeology* 9, 3 (1990), 263-297.

E. Pernicka, Th. Rehren, and S. Schmitt-Strecker 1998. "Late Uruk Silver Production by Cupellation at Habuba Kabira, Syria", in: in: Th. Rehren, A. Hauptmann, J. Muhly (eds), *Metallurgica Antiqua, Bochum*, 1998, 123-134 [Der Anschnitt 8].

E. Pernicka, C. Eibner, Ö. Öztunali, G.A. Wagner, Early Bronze Age metallurgy in the north-east Aegean, in: G.A. Wagner, E. Pernicka, and H.-P. Uerpmann (eds.), *Troia and the Troad*, Berlin/ Heidelberg, 2003, 143-172.

E. Pernicka, M. Momenzadeh, A. Vatandoust, K. Adam, M. Böhme, Z. Hezarkhani, N. Nezafati, M. Schreiner, B. Winterholler, «Archaeometallurgical Research on the Western Central Iranian Plateau», in: A. Vatandoust, H. Parzinger and B. Helwing (eds), *Early Mining and Metallurgy on the Western Central Iranian Plateau*, Mainz, 2011, 633-687.

Th. Rehren, L. Boscher, and E. Pernicka, "Large scale smelting of speiss and arsenical copper at Early Bronze Age Arisman, Iran", *Journal of Archaeological Science* 39, 6 (2012), 1717-1727.

C. Renfrew, "Cycladic metallurgy and the Aegean Early Bronze Age", *American Journal of Archaeology* 71, 1 (1967), 1-20.

C. Renfrew, *The Emergence of Civilisation*, London, 1972.

C. Renfrew, "Speculations on the use of Early Cycladic sculpture", in: J. L. Fitton (ed.), *Cycladica: Studies in Memory of N. P. Goulandris*, London, 1984, 24-30.

C. Renfrew, *The Cycladic Spirit: Masterpieces from the Nicholas P. Goulandris Foundation*. London, 1991.

C. Renfrew, "The Keros Hoard: Remaining Questions", *American Journal of Archaeology* 112 2 (2008), 295-298.

C. Renfrew, "The sanctuary at Keros: questions of materiality and monumentality", *Proceedings of the British Academy* 1, (2013), 187-212.

C. Renfrew, "A note on earlier work at the Special Deposit North", in: C. Renfrew, O. Philaniotou, N. Brodie, G. Gavalas, and M.J. Boyd (eds), *The Sanctuary on Keros and the Origins of Aegean Ritual Practice, Vol. II. Kavos and the Special Deposits*, Cambridge, 2015, 399-411 [*McDonald Institute Monographs*].

C. Renfrew, C. Doumas, L. Marangou, and G. Gavalas (eds), *Keros, Dhaskalio Kavos: The Investigations of 1987-88*, Cambridge, 2007 [*McDonald Institute Monographs*].

C. Renfrew, M.J. Boyd, and C. Bronk Ramsey, "The oldest maritime sanctuary? Dating the sanctuary at Keros and the Cycladic Early Bronze Age", *Antiquity* 86, 2012, 144-160.

C. Royd, O. Philaniotou, N. Brodie, G. Gavalas, and M.J. Boyd (eds), *The Sanctuary on Keros and the Origins of Aegean Ritual Practice, Vol. I. The Settlement at Dhaskalio*, Cambridge, 2013 [*McDonald Institute Monographs*].

C. Renfrew, O. Philaniotou, N. Brodie, G. Gavalas, and M.J. Boyd (eds), *The Sanctuary on Keros and the Origins of Aegean Ritual Practice, Vol. II. Kavos and the Special Deposits*, Cambridge, 2015, [*McDonald Institute Monographs*].

C. Renfrew, O. Philaniotou, N. Brodie, G. Gavalas, and M.J. Boyd (eds), *The Sanctuary on Keros and the Origins of Aegean Ritual Practice: the excavations of 2006-2008, Vol. III. The Marble Finds from Kavos and Archaeology of Ritual*, Cambridge, 2018a, 501-532 [*McDonald Institute Monographs*].

C. Royd, M.J. Boyd, and E. Margaritis, "Interdisciplinary approaches to the prehistory of Keros", *Archaeological Reports* 64, 2018b, 67-84.

P. Sotirakopoulou, *The "Keros Hoard" Myth or Reality?: Searching for the Lost Pieces of a Puzzle*, Athens, 2005.

P. Sotirakopoulou, "The Keros Hoard: Some Further Discussion", *American Journal of Archaeology* 112, 2 (2008), 279-294.

P. Sotirakopoulou, *The Sanctuary on Keros and the Origins of Aegean Ritual Practice Vol. IV. The Pottery from Dhaskalio*, Cambridge, 2016 [*McDonald Institute Monographs*].

P. Sotirakopoulou, *The Sanctuary on Keros and the Origins of Aegean Ritual Practice: the excavations of 2006-2008, Vol. V: The Pottery from Kavos*, Cambridge, forthcoming [*McDonald Institute Monographs*].

Z.A. Stos-Gale, N.H. Gale, and G.R. Gilmore, "Early Bronze Age Trojan metal sources and Anatolians in the Cyclades", *Oxford Journal of Archaeology* 3, 3 (1984), 23-43.

Z.A. Stos-Gale, "Bronze Age Metal Sources and the Movement of Metals between the Aegean and Anatolia", in: M. Bartelheim, B. Horejs, and R. Krauss (eds), *Von Baden bis Troia Ressourcennutzung, Metallurgie und Wissenstransfer eine Jubiläumsschrift für Ernst Pernicka*, Rahden/Westf., 2016, 375-398.

C.P. Thornton, C.P., Th. Rehren, and V.C. Pigott, "The production of speiss (iron arsenide) during the Early Bronze Age in Iran", *Journal of Archaeological Science* 36, 2 (2009), 308-316.

G.A. Wagner, and G. Weisgerber, *Silber, Blei und Gold auf Siphnos*, Bochum, 1985 [*Der Anschnitt* 3].

T.M. Whitelaw, "The Objectives and Methods of the 1987 Surface Survey at Dhaskalio Kavos", in: C. Renfrew, C. Doumas, L. Marangou, and G. Gavalas (eds), *Keros, Dhaskalio Kavos: The Investigations of 1987-88*, Cambridge, 2007, 39-78 [*McDonald Institute Monographs*].

P. Zapheiropoulou, "Κυκλάδες: ανασκαφικαί έρευναι-περιοδείαι: Κέρος", *Archeologikon Deltion (Chronika)* 23, (1968a), 381-3.

P. Zapheiropoulou, "Cycladic finds from Keros", *Athens Annals of Archaeology* 1, (1968b), 97-100.

P. Zapheiropoulou, Όστρακα εκ Κέρου, *Athens Annuals of Archaeology* 8, (1975), 79-85.

P. Zapheiropoulou, C. Doumas, and C. Renfrew, "Earlier work", in: C. Renfrew, C. Doumas, L. Marangou, and G. Gavalas (eds), *Keros, Dhaskalio Kavos: The Investigations of 1987-88*, Cambridge, 2007, 17-38 [*McDonald Institute Monographs*].

3. THE DEVELOPMENT OF COPPER-BASED METALLURGY IN THE BRONZE AGE OF THE IRANIAN PLATEAU: TRANSITION FROM ARSENICAL COPPER TO TIN BRONZE DURING THE THIRD AND SECOND MILLENNIUM BC

Omid Oudbashi and Mathias Mehofer

Abstract

Despite the importance of the metallurgy of copper alloys during the Bronze Age of the Iranian Plateau, there is limited systematic scientific data about the copper-based metallurgy and alloying processes in this period. Nevertheless, it is obvious that the tin bronze technology occurred widely through the Middle and Late Bronze Age (ca. 2500 to 1500 BC). In this paper, a brief description of the analytical results of copper-based objects are presented and analysed based on the multianalytical methods to show the emergence and development of tin bronze during the Bronze Age of Iran. The results showed that tin bronze occurred during the EBA in western Iran while copper and arsenical copper were the main materials used in other regions of the Iranian Plateau, especially central and eastern parts. Furthermore, it is worth noting that the spread of tin bronze metallurgy occurred more during the MBA and LBA. Accordingly, it can be concluded that although tin bronze metallurgy started already during the EBA in the western Iranian Plateau, it just became a commonplace metal during the second millennium BC as the dominant material in the prehistoric metallurgy of the Plateau.

Keywords

Iranian Plateau, Bronze Age, Archaeometallurgy, Arsenical Copper, Tin Bronze

Introduction

Generally speaking, the first evidence for tin bronze metallurgy can already be observed during the late fourth/early third millennium BC (ca. Early Bronze Age) in the Near East and especially in the Iranian Plateau, even though this technology was not widespread until the second millennium BC (Late Bronze Age and Iron Age).[1] From the early evidences of tin bronze in Western Iran (Luristan), it spread to other regions of Iran in later times.[2] The current information about this is based on some archaeometallurgical studies performed on prehistoric objects dated to the third and second millennium BC. In fact, it is more than necessary to initiate and develop archaeometallurgical research on the emergence and spread of tin bronze metallurgy to fully understand the archaeometallurgical activities from the Bronze Age to the Iron Age.

Early dated evidence for the usage of tin bronze was discovered in the Early and Middle Bronze Age sequences of Western Iranian archaeological sites like e.g. Kalleh Nisar, Bani Surmeh, Susa, Deh Dumen and Godin Tepe (Godin Tappeh). Though, besides many copper and arsenical copper artefacts, the number of low-tin bronze objects is limited.[3] At the end of the third millennium BC the situation changed and arsenical copper/arsenical bronze is regularly replaced by tin bronze in different regions, as it can be pointed out in Susa or Malyan.[4] In general, tin bronze became the commonplace used metal during the Late Bronze Age in most parts of the Plateau.[5] Nevertheless, tin bronze objects are rare in the north-eastern and eastern Iran even during the Late Bronze Age as is demonstrated from important archaeological sites such as Tepe Yahya

1 Maddin *et al.* 1999; Moorey 1988; Tylecote 1992; Avilova 2009; Oudbashi *et al.* 2012; Pigott 2004; Thornton 2009a; Mehofer 2014; Oudbashi 2019.

2 Helwing 2013; Pigott 2004.

3 Fleming *et al.* 2005; Nezafati 2006; Oudbashi *et al.* 2016; Frame 2010.

4 De Ryck *et al.* 2003; Pigott *et al.* 2003a.

5 Pigott 2004; Oudbashi *et al.* 2012; Oudbashi 2019.

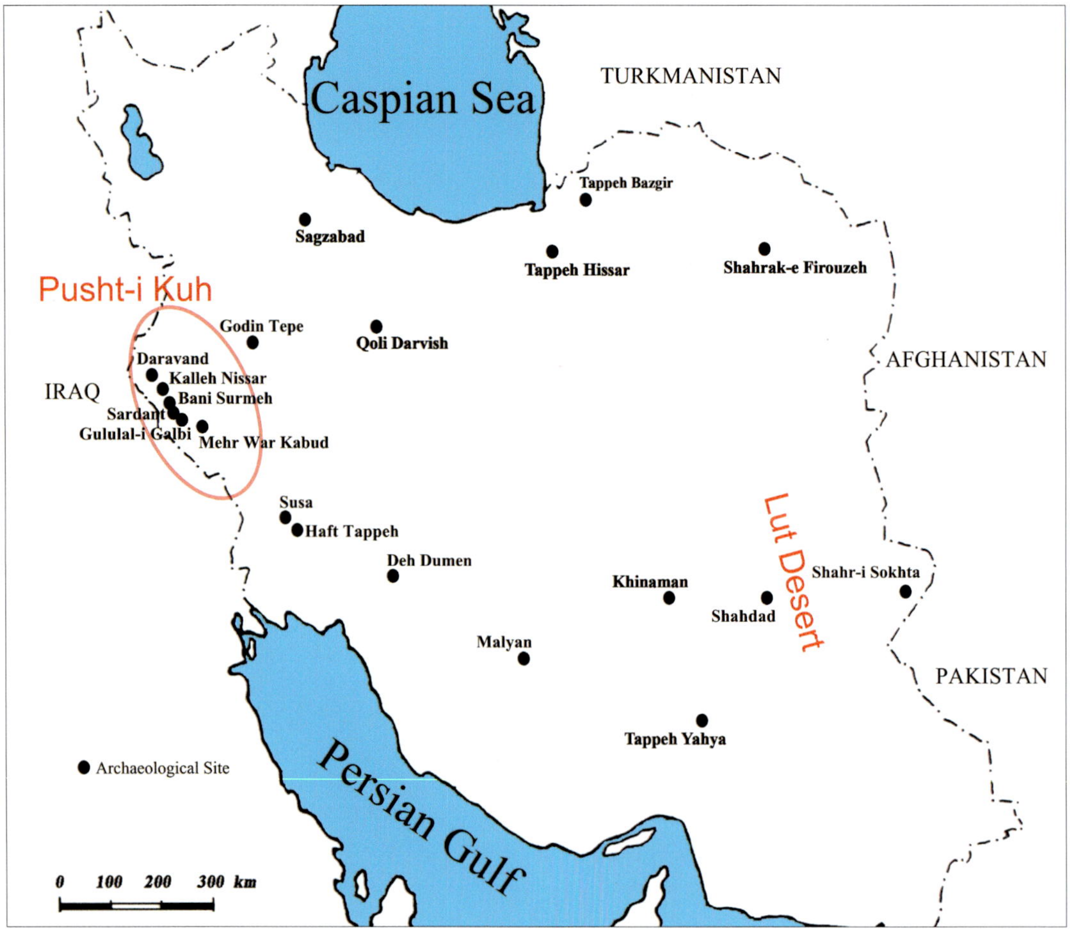

Fig. 1. Map of Iran and name and location of the archaeological sites mentioned in the text.

(Tappeh Yahya), Tepe Hissar (Tappeh Hissar), Shahr-i Sokhta and Shahdad.[6]

Apart from the available data, the development of tin bronze metallurgy in the Iranian Plateau is not well-known and it is necessary to develop archaeometallurgical studies to identify the process of formation of tin bronze metallurgy and transition from Cu-As to Cu-Sn technology in this region. In this paper, a brief review and comparative explanation on available analytical data of copper-based metallurgy during the Bronze Age of Iran is reported. It includes the previously as well as recently published works presenting results of chemical composition of archaeological objects and tries to show the metallurgical phenomena occurred based on different regions of Iran (Fig. 1).

6 Thornton *et al.* 2002; Thornton *et al.* 2004; Meier 2011; Hauptmann *et al.* 2003.

The Old Problem

For the Iranian Plateau the Bronze Age is defined as a period which lasted from ca. 3000-1500 BC. It indicates certain developments in culture, economy, trade and industry. Large centers or cities appeared during this period at the Plateau such as Shahr-i Sokhta, Shahdad and Susa, and evidence of long-distance trade and exchange is apparently visible in this time. Although, this period is known as the Bronze Age (according to the periodization of prehistory worldwide) it is more about cultural and economic changes than a metallurgical innovation in the Iranian Plateau. As mentioned above, results of analytical studies on copper-based objects from the Bronze Age of Iran shows that a variety of copper alloy compositions were in use. According to the available information, the aspects of copper-based metallurgy during the third and second millennium BC is identified partially and many questions are now arising, for instance:

- What is the main copper-based technology during the Bronze Age of the Plateau?
- When did the tin bronze technology emerge, spread and develop?
- What was the first alloying process for tin bronze production in the early stages (e.g. mixing metallic copper with tin ore or mixing metallic copper with metallic tin?)?
- What were the alloying process used after spread of tin bronze metallurgy in the Plateau?
- Where can the ore resources (copper and especially tin) be located that were used to produce tin bronze objects?
- Where are the workshops/industrial places in which the objects are produced or the metal is smelted?
- and ...

These questions (and more) show that the emergence and spread of tin bronze metallurgy in the study region is not fully understood. It is necessary to develop archaeological and archaeometallurgical research in this field.

As a starting point, a short-time project was planned. Within an ongoing cooperation project between the Art University of Isfahan and the Vienna Institute for Archaeological Science, University of Vienna, the emergence and spread of tin bronze metallurgy in the Iranian Plateau during the Bronze Age (3rd-2nd millennium BCE) was investigated. The project included application of analytical studies to identify the metallurgical processes used in the production of the Bronze Age (third and second millennium BC) objects from Iran as well as a re-consideration of the available information about the Bronze Age copper-based metallurgy in this region. The results of the analytical works are under evaluation and publication while some of them are already published. Furthermore, a brief explanation of available and recent analytical data will be interpreted here to show the achievements of the project until now.

Copper-based Metallurgy in the Bronze Age: Third Millennium BC

Although the starting of the Bronze Age is dated back to the beginning of third millennium BC, yet this period is started in partially different times in the different regions of the Iranian Plateau. Thus, it can be seen that the Bronze Age is dated back to the last centuries of the fourth millennium BC. However, analytical studies on EBA and MBA archaeological copper alloys from the Plateau revealed very significant aspects.

Western Iran

Luristan (western Iran) is known as the first region in which tin bronze emerged and was used for small objects. Analysis of a significant number of copper-based objects excavated at the Early and Middle Bronze Age sites of Pusht-i Kuh (Luristan) showed that among 87 objects dated back to third millennium BC, 43 objects are made of tin bronze while 27 objects are made of arsenical copper and copper. Also, the earliest evidence of application of tin bronze to produce objects is observed in the Kalleh Nisar Bronze Age graveyard in Luristan.[7] On the other hand, analysis of 38 copper-based objects dated back to late 4th-3rd millennium from the Godin Tepe archaeological site, north of Luristan, shows that only 7 objects are made of tin bronze while others are copper and arsenical copper. The tin bronze objects from Godin Tepe are from different levels of the site including VI:1, IV, III, III:6, III:5 and III:4.[8]

Southwestern Iran

The archaeological site of Susa, located in southwestern Iran, is a well-known site in ancient Iran. It shows various levels of settlements from prehistoric to Islamic periods in the Susiana plain. A large-scale archaeometallurgical study was performed on copper-based objects dated back to 6th to 3rd millennium BC.[9] This analyzed collection consists of 280 objects from Susa I to Susa V levels. The analyses showed that the composition of objects, which are dated back to the third millennium BC (Susa IIIa, IIIb, IVa, IVb and V), include tin bronze and arsenical copper, although low arsenical copper was used during the entire time span (I-V levels; ca. 5250-2000 BC). Objects with high arsenic content start to appear in period Susa II-IIIa (ca. 3700-2950 BC). Furthermore, the use of tin bronze is approved since the end of Susa IVa (ca. 2350 BC) and continued to IVb and V levels (ca. 2350-2000 BC).[10]

The recently excavated Bronze Age graveyard in southwestern Iran, Deh Dumen, is located in the Zagros mountain at the west of Susiana Plain. The archaeological and analytical works revealed that the graveyard is dated back to the third and second

7 Fleming *et al.* 2005; Begemann *et al.* 2008.

8 Frame 2010.

9 Malfoy and Menu 1987; De Ryck, Adriens and Adams 2005.

10 De Ryck, Adriens and Adams 2005.

millennium BC comparable with Bronze Age and Early Iron Age.[11] The copper-based objects are commonplace in all graves and various objects with different typologies were discovered at this site. Analysis of ten samples from eight thin sheet vessels dated back to the mid of third millennium BC showed that all the bodies of vessels are made of tin bronze while two metallic massive places inside the base of two vessels are made of low arsenical copper.[12]

Eastern Iran

Although archaeological excavations in eastern Iran in prehistoric period are significantly performed yet analytical data about copper-based metallurgy, in particular during the Bronze Age, is not considerable. Nevertheless, available data presents a general shape of copper-based metallurgy during the third and second millennium BC in the region.

Shahdad is one of the important eastern archaeological sites that is excavated extensively and shows a big Bronze Age settlement beside the Lut dessert. The settlement is dated to the third millennium BC and provides significant evidence for copper-based metallurgy, metal objects, metallurgical materials as well as a workshop. Unfortunately, comprehensive archaeometallurgical examinations were not undertaken on the site and only some results are published. The analytical data of some objects (23 objects) and slag fragments from the site show that the main copper-based technology in this site is copper smelting and production of objects made of copper and low arsenical copper.[13] Furthermore, eighteen analyzed objects from Khenaman or Khinaman, a Bronze Age cemetery located in the Lut Desert, which is related to the Shahdad civilization, showed that three tin-containing copper objects is observed together with arsenical copper and copper artefacts.[14]

Tappeh Yahya (Tepe Yahya), another important prehistoric site in southwestern Iran (Kerman), is dated back from the 6th to the 1st millennium BC (Neolithic era to Iron Age). Analysis of 35 objects dated back to third millennium BC (IVC-IVB levels) shows that copper objects (and arsenical copper in some cases) are the main products.[15] On the other hand, three other analyzed objects from this site (third millennium BC) are made of copper and leaded copper.[16]

The well-known Bronze Age site of Shahr-i Sokhta in Sistan, dated from late 4th to 3rd millennium BC, is a large-scale urban settlement from the Bronze Are of eastern Iran. In fact, the site can be mentioned as one of the biggest Bronze Age urban sites in the ancient world. The settlement is extensively excavated and evidence of copper smelting as well as copper-based objects are discovered. Results of analysis of ores, slags and metallic fragments from the Early Bronze Age (ca. 2700-2500 BC) revealed that co-smelting of complex copper ores (oxidic and sulfidic) in crucible was the main metallurgical process in that period. This process was performed to produce copper and arsenical copper objects.[17]

Northcentral Iran

The only archaeometallurgical study focusing on a site located in the region was performed on objects discovered by E. F. Schmidt during 1930s from Tappeh Hissar.[18] The dating of the site (based on archaeological finds) shows a sequence through the 4th to 3rd millennium BC (about 2000 years). A palace from the Sasanian period (ca. 3rd to 7th centuries AD) was also unearthed.[19] The results of the chemical analysis of 228 copper-based objects from I to IIIC periods (4th and 3rd millennium BC) revealed that the two major alloying elements in the Tappeh Hissar copper-based metal artifacts are lead and arsenic (accidentally or intentionally). In fact, the main copper alloys used during the third millennium BC are leaded copper and arsenical copper (IIIA to IIIC periods). Only one object with significant amount of tin (3.6 wt.%) dates back to IIIC period or late 3rd -early 2nd millennium BC. Also, some rare evidence of zinc-containing copper objects is reported from this prehistoric site.[20] Consequently, results of analytical studies on a huge collection of copper-based objects from this site showed no evidence of tin bronze metallurgy in this area and revealed the accidental and intentional alloying of copper with lead and arsenic during Chalcolithic and Bronze Age.

11 Oudbashi *et al.* 2016; Miszkiewicz *et al.* 2022.

12 Oudbashi *et al.* 2016; Oudbashi *et al.* 2017.

13 Meier 2011; 2015; Oudbashi *et al.* 2017.

14 Curtis 1988; Maxwell-Hyslop 1988.

15 Thornton *et al.* 2002.

16 Meier 2011.

17 Hauptmann *et al.* 2003.

18 Thornton 2009b.

19 Schmidt 1937.

20 Thornton 2009b; Thornton and Rehren 2009.

Southcentral Iran

Excavations at Malyan, an important archeological site in the Kur River Basin in central Fars (southcentral Iran), identified as ancient Anshan, the highland capital of Elam, revealed some evidence of metallurgical activities dated back to Banesh culture (ca. 3400-2800 BC). Evidence for copper smelting come from Operation Banesh-TUV, Levels 2-3, where slag, bits of metal, and furnace fragments were discovered in domestic contexts, while some buildings (Operation Banesh-ABC) also yielded some evidence for copper smelting.[21] According to analyses the copper contains arsenic, lead and, in one case, antimony.[22]

Copper-based Metallurgy in the Bronze Age: Second Millennium BC

Western Iran

In contrast with the third millennium BC, copper-based metallurgy during the first half of the second millennium BC has not been studied widely in western Iran and Luristan. The majority of analyzed objects from Godin Tepe were dated back to the third millennium BC, the analysis of 18 copper-based objects from second millennium BC of Godin Tepe (III:2 and Post III:2 levels) also showed that 13 objects are made of tin bronze while others are made of arsenical and impure copper.[23] It shows that tin bronze has been used in more content during the first half of the second millennium BC in Godin Tepe.

Southwestern Iran

Analytical studies on objects from Susa are limited to the end of the third millennium BC and no objects from second millennium BC is analyzed from this site.[24] Also, there is no published data about objects dated back to Late Bronze Age. Only one analytical work was undertaken on the metallurgical materials and metal objects from the middle Elamite site of Haft Tappeh near Susa that is dated back to the 14th century BC, a bit after the end of the Bronze Age in the Iranian Plateau. The site that is known as the capital of the Elamite king, Tepti Ahar, was excavated extensively and evidence for copper production and metallurgy including furnace, crucibles, ingots, slags and objects are discovered in its industrial part.[25] A multianalytical study on metallic objects, ingots and slags revealed that the metallurgical activity in Haft Tappeh was small-scale production of copper and tin bronze objects from sulphidic raw materials by partial smelting of the ores leading to the production of copper ingots with high amounts of impurities. The product of the first stage was the copper matte that was refined to obtain copper ingots with lower impurities. The production of tin bronze alongside impure copper indicates the application of an uncontrolled alloying process.[26]

Southcentral Iran

Some other objects from Malyan dated back to the Kaftari culture (ca. 2200-1600 BC) are analyzed to identify the chemical composition of copper-based materials.[27] PIXE analysis of ten copper-based objects revealed that six objects are made of tin bronze and tin is detected in variable amounts while two objects are made of arsenical copper and others are impure copper objects. In fact, tin bronze objects from Kaftari Phase represent the earliest evidence of bronze metallurgy in southcentral Iran.

Northcentral Iran

Archaeometallurgical studies on Late Bronze Age materials coming from northcentral Iran were more performed during the last decades. Archaeological excavations in the northcentral Iran unearthed small- and large-scale copper-based collections such as excavations in Sagzabad (Qazvin Plain), Qoil Darvish (Near Qom) and Tappeh Bazgir (near Gorgan).

Sagzabad, a well-known archaeological site in Qazvin Plain among with Qabristan and Tappeh Zagheh, is excavated for about 50 years and a large settlement and its large graveyard dated back to Late Bronze Age and Iron Age is identified.[28] Although, two archaeometallurgical studies are undertaken on objects from Sagzabad, only four objects were belonging to the first half of the second millennium BC (Late Bronze Age).[29] Chemical analysis of objects showed that three objects from the Late Bronze Age are made of tin bronze and one object is made of impure

21 Abdi 2004.

22 Pigott *et al.* 2003b.

23 Frame 2010.

24 De Ryck *et al.* 2003.

25 Negahban 1991.

26 Oudbashi and Mishamastnehi 2020 ; Oudbashi *et al.* 2019.

27 Pigott *et al.* 2003a.

28 Fazeli *et al.* 2005; Negahban 1977.

29 Oudbashi *et al.* 2021a; Mortazavi *et al.* 2011.

copper. On the other hand, the Iron Age objects are of impure copper, tin bronze, and in one case, arsenical copper. It reveals that tin bronze metallurgy has been started at the late Bronze Age and is continued to Iron Age but copper (and arsenical copper) were also used during the Iron Age.

During a construction operation in 2006, a large scale copper-based collection was unearthed in Tappeh Bazgir of Minoodasht, near Gorgan. The huge amount of copper-based objects discovered from the site, including more than 700 objects with varying dimensions, forms one of the important archaeometallurgical ensembles in northern Iran dating back to 1800-1700 BC.[30] 171 metallic objects from this site were analyzed previously and revealed that copper-arsenic alloy has been used to produce these artefacts, and only two buttons are made of tin bronze.[31]

Qoli Darvish, a large prehistoric settlement near Qom, is located at the eastern edge of the central dessert of Iran at the southern borders of the northcentral Iran and central Iran. The site shows a sequence settlement from Chalcolithic to Late Bronze Age. Analytical works on four objects besides some archaeometallurgical materials showed that only one object is made of tin bronze while the remaining are made of copper and arsenical copper.[32] Although, the small number of analyzed object does not provide a complete picture of the copper-based metallurgy in this site, analysis of slag and other archaeometallurgical materials shows that copper production was carried out on the site.

Northeastern Iran

As observed in the third millennium BC, early metal production in northeastern Iran is not studied and identified well. Nevertheless, recent archaeological excavations in Neyshabur (Nishabur) caused to discover a Late Bronze Age settlement. Evidence of an ancient settlement in the area, namely Shahrak-e Firouzeh, was discovered during construction activities in the modern city of Neyshabur. The archaeological findings include a settlement with industrial workshops and a cemetery that, based on the type of constructions and the objects and grave goods, the site has been dated back to the Middle and Late Bronze Age; during this time, it has also had relations with the important cultures of the second millennium BC in Great Khorasan and Central Asia, including Bactro-Margiana Archaeological Complex/ Great Khorasan Civilization (BMAC/GKC - ca. 2100-1700 BC).[33]

According to chemical analysis of 23 objects, it can be stated that unalloyed and arsenical copper have been used to make different metallic objects discovered in Shahrak-e Firouzeh LBA settlement. The interesting aspect of this study is that there is no evidence whatsoever of using tin bronze (only two cases with low-amount of tin) in the site.[34] However, this is the only archaeometallurgical study performed on Bronze Age objects from northeastern Iran and it needs to develop analytical works on excavated copper-based objects from the region.

Eastern Iran

The IVA level of Tappeh Yahya, dated back to ca. 1800-1400 BC, also shows considerable amounts of copper-based objects; 34 object from the earlier levels were analyzed among others.[35] This level shows interesting aspects of copper-based metallurgy in eastern Iranian Plateau: six object have significant amounts of tin and three objects are made of copper-zinc. Others show copper and arsenical copper metallurgy. These tin bronze objects – besides objects found from Khinaman- can be introduced as only tin bronze objects found in eastern Iran during the Bronze Age. In addition, the presence of objects with high concentration of zinc is another interesting aspect in Tappeh Yahya, although it is not possible to attribute them to an intentional alloying process.

Tin Bronze or Arsenical Copper: A General View

The problem of emergence of tin bronze metallurgy during the Bronze Age of the Iranian Plateau, can be partially resolved by looking at the available data of analytical works, although these studies are not comprehensive enough to show different aspects of copper-based metallurgy in this rich region.
Western and southwestern Iran can be considered as the earliest regions in which tin bronze has emerged during the Early and Middle Bronze Age. Although the dating of objects from Pusht-i Kuh needs to be reconsidered, results of analytical studies show the emergence of tin bronze metallurgy besides copper

30 Nokandeh *et al.* 2006.

31 Lorenz 2008.

32 Nezafati and Stöllner 2017.

33 Basafa 2020.

34 Oudbashi *et al.* 2021b.

35 Thornton *et al.* 2002.

and arsenical copper in Luristan between about 3000-2500 BC. On the other hand, analytical works on western and southwestern Iranian sites show that tin bronze technology has emerged during the second half of the third millennium BC, as was observed in Susa, Deh Dumen and Godin Tepe. Of course, this appeared among with the existing technology of arsenical copper and copper, and tin bronze has not completely replaced the older alloy. Also, tin bronze, arsenical copper and copper objects are observed together in the second millennium BC in western Iran (Godin Tepe and Haft Tappeh), showing that the copper and arsenical copper metallurgy has not been completely with tin bronze during the late Bronze Age, and even early Iron Age.

Southcentral Iran is another region that has been studied to find evidence of copper-based metallurgy during the Bronze Age, although there is only one Group of analytical data from Malyan. This data obviously revealed that arsenical copper has been the main metallurgical activity in Malyan during the late Chalcolithic and Early Bronze Age (Banesh culture) while tin bronze appeared in the Middle and Late Bronze Age (Kaftari culture). Of course, arsenical copper has been presented during the last period.
Despite the large-scale excavations in the Bronze Age sites of eastern Iran, the available analytical data about their metallurgy and metallic objects is not comprehensive. Nevertheless, the available data present an acceptable view about metallurgical changes during this period. Arsenical copper and copper metallurgy has been the main activity during the third millennium BC in eastern Iran and surrounding the Lut dessert and only few tin bronze objects are discovered from Khinaman, near Shahdad. Furthermore, this is continued during the late Bronze Age and only few (6) tin bronze objects are identified among a large group of copper, leaded and arsenical copper objects in Tappeh Yahya IVA level. It shows that tin bronze has not emerged during the Bronze Age in eastern Iran and these rare examples should be considered as exceptions due to some trade and exchange connections in the Bronze Age of Iran.
The northern part of Iran has a rich history during the Bronze Age and numerous archeological sites are excavated and studied. Nevertheless, available archaeometallurgical data are not enough to find a specific technological pattern about copper-based metallurgy. Therefore, it is obvious that arsenical and leaded copper were used during the third millennium BC and apart from one tin bronze object from Tappeh Hissar, no evidence of tin bronze metallurgy has been observed there. On the other hand, arsenical copper, copper and tin bronze have been observed among the analyzed objects from second millennium BC in northcentral Iran. Although tin bronzes are only observed in limited number in Sagzabad and Qoli Darvish. It shows that arsenical copper has been commonplace in northcentral Iran, as also is observed in the only analyzed groups of Late Bronze Age objects from Shahrak-e Firouzeh, northeastern Iran, in which some low-tin bronzes are detected among copper and arsenical copper objects.

Figure 2 shows the map of Iran and a schematic figure of developments, changes and evolutions of copper-based metallurgy during the Bronze Age of Iran. According to presented data in Table 1 and Figure 2, it can be derived that tin bronze metallurgy has emerged during the third millennium BC in western and southwestern Iran while copper and arsenical copper were used extensively in the rest of Iran. In fact, tin bronze metallurgy has emerged during the Bronze Age only in the western part of the Plateau while copper and arsenical copper were also used in this region. Tin bronze objects emerged in other parts of the Plateau during the second millennium BC (Late Bronze Age) such as northcentral, southcentral as well as northeastern Iran. It is worth noting that tin bronze metallurgy has not emerged during the Bronze Age in eastern Iran including surrounding of the Lut dessert, and very few examples of tin bronze in this region should be mentioned as exceptions.
According to the map in Figure 2, Iran can be divided to three regions based on results of the analytical studies on copper-based objects:

- Region I, including western and southwestern Iran in which tin bronze has emerged and was used during the third millennium BC.
- Region II, including northcentral, southcentral and northeastern Iran in which no apparent evidence of tin bronze has been detected during the third millennium BC while tin bronze has emerged in some significant amounts during the second millennium BC.
- Region III, including eastern Iran and Lut desert, in which tin bronze has not emerged and used during the Bronze Age, apart from some limited exceptions.

Consequently, although the Bronze Age (ca. 3000-1500 BC) can be assumed as the period of emergence and spread of tin bronze metallurgy in the Iranian Plateau, this technology was widespread and commonplace during the Iron Age, as is observed from the analytical works on this period.[36] Achieving a

36 e.g. Oudbashi *et al.* 2021c; Oudbashi and Hessari 2017;

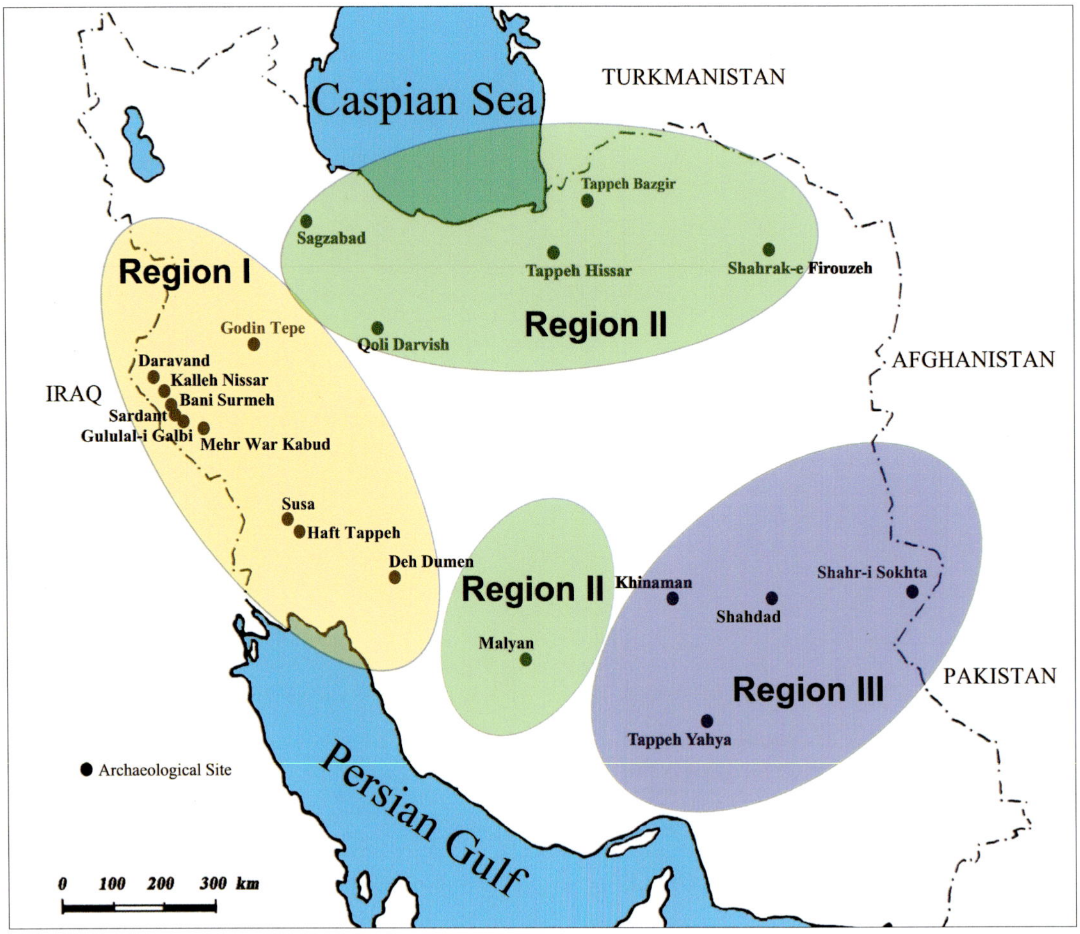

Fig. 2. Three categorized regions based on results of analysis of copper-based objects, Region I: western and southwestern Iran, Region II: Northcentral, southcentral and north eastern Iran, Region III: eastern Iran.

comprehensive image about copper-based metallurgy and spreading tin bronze technology in the Iranian Plateau during the Bronze Age needs to develop archeological and archaeometallurgical studies to clarify some unclear aspects of this technological change, from metallurgical and provenance viewpoints.

Conclusion

Metallurgy and production of tin bronze objects in the Iranian Plateau is not well explored until now, though, some analytical works have been performed during the last decades to identify the production technology as well as raw material achievements. Even though, from a chronological point of view, the Bronze Age in the Iranian Plateau starts at the beginning of the third millennium BC, the emergence of tin bronze metallurgy does not start simultaneously. The results of analytical works provided information that tin bronze emerged during the third millennium BC (Early and Middle Bronze Age) in western and southwestern Iran (region I) while arsenical copper and copper have been commonplace in these regions at the same time. On the other hand, there is no obvious evidence of tin bronze use in other parts of the Plateau during the third millennium BC. This new alloy appeared in central regions of Iranian Plateau (Region II) at the end of the third millennium as well as early second millennium BC, although arsenical and leaded as well as impure copper have been used extensively in these regions. Lastly, eastern Iran (Region III) can be mentioned as a no tin bronze region during the Bronze Age. It shows that Bronze Age should be declared as the period of starting tin bronze metallurgy and continuing arsenical copper and copper metallurgy in the history and archaeology of the Iranian Plateau. Nevertheless, further archaeological and archaeometallurgical studies are necessary to find more detailed information about the emergence and spread of tin bronze metallurgy during the prehistoric period of Iran.

Oudbashi *et al.* 2018; Fleming *et al.* 2006.

<table>
<tr><th colspan="2">Date (BC)</th><th>3100</th><th>3000</th><th>2900</th><th>2800</th><th>2700</th><th>2600</th><th>2500</th><th>2400</th><th>2300</th><th>2200</th><th>2100</th><th>2000</th><th>1900</th><th>1800</th><th>1700</th><th>1600</th><th>1500</th><th>1400</th><th>1300</th></tr>
<tr><td rowspan="2">Western Iran</td><td></td><td colspan="11">Pusht-i Kuh</td><td></td><td></td><td></td><td></td><td></td><td></td><td></td><td></td></tr>
<tr><td></td><td></td><td colspan="17">Godin Tepe</td><td></td></tr>
<tr><td colspan="2">Metallurgy</td><td colspan="11">Copper and Arsenical Copper, Tin Bronze appeared</td><td colspan="7">Tin Bronze widespread, Copper and Arsenical Copper also used</td><td></td></tr>
<tr><td rowspan="3">Southwestern Iran</td><td></td><td colspan="11">Susa</td><td></td><td></td><td></td><td></td><td></td><td></td><td></td><td></td></tr>
<tr><td></td><td></td><td></td><td></td><td></td><td></td><td colspan="6">Deh Dumen</td><td></td><td></td><td></td><td></td><td></td><td></td><td></td><td></td></tr>
<tr><td></td><td></td><td></td><td></td><td></td><td></td><td></td><td></td><td></td><td></td><td></td><td></td><td></td><td></td><td></td><td></td><td></td><td colspan="3">Haft Tappeh</td></tr>
<tr><td colspan="2">Metallurgy</td><td colspan="11">Copper and arsenical copper, Tin Bronze appeared about 2500 BC</td><td colspan="8">Copper and Tin Bronze</td></tr>
<tr><td rowspan="4">Northcentral Iran</td><td></td><td colspan="15">Tappeh Hissar</td><td></td><td></td><td></td><td></td></tr>
<tr><td></td><td></td><td></td><td></td><td></td><td></td><td></td><td></td><td></td><td></td><td></td><td></td><td colspan="6">Sagzabad</td><td></td><td></td></tr>
<tr><td></td><td></td><td colspan="16">Qoil Darvish</td><td></td><td></td></tr>
<tr><td></td><td></td><td></td><td></td><td></td><td></td><td></td><td></td><td></td><td></td><td></td><td></td><td></td><td colspan="3">Tappeh Bazgir</td><td></td><td></td><td></td><td></td></tr>
<tr><td colspan="2">Metallurgy</td><td colspan="11">Arsenical Copper, Leaded Copper, No evidence of Tin Copper</td><td colspan="8">Arsenical Copper and Tin Bronze</td></tr>
<tr><td>Southcentral Iran</td><td></td><td colspan="4">Malyan-Banesh</td><td></td><td></td><td></td><td></td><td></td><td colspan="7">Malyan-Kaftari</td><td></td><td></td><td></td></tr>
<tr><td colspan="2">Metallurgy</td><td colspan="11">Arsenical Copper and leaded Copper</td><td colspan="8">Copper and Arsenical Copper, Tin Bronze</td></tr>
<tr><td rowspan="4">Eastern Iran</td><td></td><td></td><td colspan="11">Shahdad</td><td></td><td></td><td></td><td></td><td></td><td></td><td></td></tr>
<tr><td></td><td></td><td colspan="11">Khinaman</td><td></td><td></td><td></td><td></td><td></td><td></td><td></td></tr>
<tr><td></td><td colspan="18">Tappeh Yahya</td><td></td></tr>
<tr><td></td><td colspan="10">Shahr-i Sokhta</td><td></td><td></td><td></td><td></td><td></td><td></td><td></td><td></td><td></td></tr>
<tr><td colspan="2">Metallurgy</td><td colspan="12">Arsenical Copper, some limited evidences of Tin Bronze</td><td colspan="7">Arsenical Copper, some evidences of Tin Bronze</td></tr>
<tr><td>Northeastern Iran</td><td></td><td></td><td></td><td></td><td></td><td></td><td></td><td></td><td></td><td></td><td></td><td></td><td colspan="4">Shahrak-e Firouzeh</td><td></td><td></td><td></td><td></td></tr>
<tr><td colspan="2">Metallurgy</td><td colspan="11"></td><td colspan="8">Arsenical Copper, low Tin Bronze</td></tr>
</table>

Table 1- A chronological chart showing the studied sites and copper-based metallurgy evolutions during the Bronze Age of Iran.

Bibliography

K. Abdi, "Craft Activities in Banesh Period Kur River Basin", in Th. Stöllner, R. Slotta, A. Vatandoust (eds), *Persia's Ancient Splendour, Mining, Handicraft and Archaeology*, Deutsches Bergbau- Museum, Bochum, 2004, 2004, 258-263.

L.I. Avilova, "Models of Metal Production in the Near East: (Chalcolithic – Middle Bronze Age), Archaeology", *Ethnology and Anthropology of Eurasia* 37, 3 (2009), 50-58.

F. Begemann, E. Haerinck, B. Overlaet, S. Schmitt-Strecker, F. Tallon, "An archaeo-metallurgical study of the early and middle bronze age in Luristan, Iran", *Iranica Antiqua* XLIII (2008), 2-66.

J. Curtis, "A reconsideration if the cemetery at Khinaman, Southeast Iran", *Iranica Antiqua* XXIII, (1988), 97-124.

I. De Ryck, A. Adriaens, F. Adams, "An Overview of Mesopotamian Bronze Metallurgy during the 3rd Millennium BC", *Journal of Cultural Heritage* 6 (2005), 261-268.

H. Fazeli, E.H. Wong, D.T. Potts, "The Qazvin Plain revisited: A reappraisal of the chronology of northwestern Central Plateau, Iran, in the 6th to 4th millennium BC", *Ancient Near Eastern Studies* 42 (2005), 3-82.

S.J. Fleming, V.C. Pigott, C.P. Swann, S.K. Nash, E. Haerinck, B. Overlaet, "The archaeometallurgy of war Kabud. Western Iran", *Iranica Antiqua* XLI (2006), 31-57.

S.J. Fleming, V.C. Pigott, C.P. Swann, S.K. Nash, "Bronze in Luristan: Preliminary Analytical Evidence from Copper/bronze Artifacts Excavated by the Belgian Mission in Iran", *Iranica Antiqua* XL (2006), 35-64.

L. Frame, "Metallurgical Investigations at Godin Tepe, Iran, Part I: The Metal Finds", *Journal of Archaeological Science* 37, (2010) 1700-1715.

A. Hauptmann, Th. Rehren, S. Schmitt-Strecker, "Early Bronze Age copper metallurgy at Shahr-i Sokhta (Iran), reconsidered", in: Th. Stöllner, G. Körlin, G. Steffens, J. Cierny (eds), *Man and Mining—Mensch und Bergbau*, Der Anschnitt, Beiheft 16, Deutsches Berbgau-Museum, Bochum, 2003, 197-213.

B. Helwing, "Early Metallurgy in Iran- an Innovative Region as Seen from the Inside", in: S. Burmeister, S. Hansen, M Kunst, N. Müller-Scheessel (eds), *Metal Matters: Innovative Technologies and Social Change in Prehistory and Antiquity*, Menschen-Kulturen-Traditionen, Forschungs Cluster 2, Band 12, DEUTSCHES ARCHAOLOGISCHES INSTITUT, Berlin, 2013, 105-136.

S. Lorenz, *Archäometrischeuntersuchungenamhortfund von Bazgir, Iran, Diplomarbeit-Magister, Institut für Archämetries*, Fakultät für Werkstoffwissenschaft und Werkstofftechnologie, TU Bergakademie Freibreg, 2008 (Unpublished).

R. Maddin, J.D. Muhly, T. Stech, "Early metalworking at Çayönü", in: A. Hauptmann, E. Pernicka, Th. Rehren, Ü Yalçin (eds), The Beginnings of Metallurgy, Ed. Der Anschnitt, Beiheft 9, Bochum, 1999, 36-44.

J.M. Malfoy, M. Menu, "LaMétallurgie du cuivre à Suse aux IVe et IIIe Millénaires: Analyses en Laboratoire", in: F. Tallon, J.M. Mafoy (eds), *Métallurgie Susienne I: 1. De la fondation de Suse au XVIIIe avant J.-C*, Notes et documents des musées de France, Paris, 1987.

K.R. Maxwell-Hyslop, "A comment on the finds from Khinaman", *Iranica Antiqua* XXIII (1988), 129-138.

M. Mehofer, "Metallurgy during the Chalcolithic and the Beginning of the Early Bronze Age in Western Anatolia", in: B. Horejs, M. Mehofer (eds), *Western Anatolia before Troy - Proto-Urbanisation in the 4th millennium BC*, Proceedings of the International Symposium held at the Kunsthistorisches Museum Wien, Vienna, Austria, 21-24 November 2012, Oriental and European Archaeology 1, Wien, 2014, 463-490.

D.M.P. Meier, "Preliminary Archaeometallurgical Investigations of Bronze Age Metal Finds from Shahdad and Tepe Yahya", *Iranian Journal of Archaeological Studies* 1, (2011), 25-34.

J.J. Miszkiewicz, T.J. Stewart, R. Naseri, A. Sołtysiak, "The lifestyles of Bronze Age Zagros highlanders at Deh Dumen, Iran: Insights from midshaft femur cross-sectional geometry and histology", *Archaeometry* 64, (2022) 1270-1287.

P.R.S. Moorey, "Early metallurgy in Mesopotamia", in R. Maddin (ed.), *The Beginning of the Use of Metals and Alloys*, The MIT Press, London, 1988. 28-33.

M. Mortazavi, A. Salehi Kakhki, M.A. Golozar, H. Tala'i, "Preliminary Metallurgical Investigation of Copper-based Artifacts at Tepe Sagzabad in Qazvin Plain, Iran (1500-800 BC)", *Iranian Journal of Archaeological Studies* 1(2) (2011), 49-59.

E.O. Negahban, *Excavation at Haft Tepe, Iran*, The University Museum of Archaeology and Anthropology, University Museum Monograph 70. University of Pennsylvania, Philadelphia, 1991.

E.O. Negahban, 1977, "Preliminary report of Qazvin expedition: Excavation of Zaghe, Qabristan, Sagzabad (1971-72)", *Marlik* 2 (1977), 26-44 (in Persian).

N. Nezafati, *Au-Sn-W-Cu-Mineralization in the Astaneh-Sarband Area, West Central Iran, including a comparison of the ores with ancient bronze artifacts from Western Asia*, PhD Dissertation, Der Geowissenschaftlichen Fakultät, Der Eberhard-Karls-Universität Tübingen, Germany, 2006 (Unpublished).

N. Nezafati, Th. Stöllner, "Economic Geology, Mining Archaeological and Archaeometric Investigations at the Veshnaveh Ancient Copper Mine, Central Iran", *Metalla* 23.2 (2017), 67-90.

O. Oudbashi, "The Bronze Metallurgy in the Iron Age Tin Bronze Production Tradition during the Iron Age in the Central Zagros", in: Y. Hassanzadeh, A.A. Vahdati, Z. Karimi (eds), *Proceedings of the International Conference on The Iron Age in Western Iran and Neighbouring Regions*, 2-3 Nov. 2019, Kurdistan University, Sanandaj, Iran, RICHT, National Museum of Iran, Kurdistan ICHHTO Publications, 2019, 201-214.

O. Oudbashi, M. Mehofer, B. Bahadori, A. Aliyari, H. Tala'i, "The Emergence and Spread of Tin Bronze Alloying in: Prehistoric Iran The LBA Metallurgy in Sagzabad, Northern Iran", in B. Török, A. Giumlia-Mair (eds), *Proceedings of the 5th Archaeometallurgy in Europe*, 19-21 June 2019 Miskolc, Editions Mergoil, 2021a, 255-269.

O. Oudbashi, M. Mehofer, H. Basafa, J. Tayyari, S. Bahadori, "Arsenical copper and bronze metallurgy during Late Bronze Age of north-eastern Iran: evidences from Shahrak-e Firouzeh archaeological site", *Archaeological and Anthropological Sciences* 12 (2021b), 231.

O. Oudbashi, F.W. Rademakers, F. Vanhaecke, P. Degryse, A. Hasanpour, S. Bahadori, "An old problem in a new light: Elemental and lead isotopic analysis of Luristan Bronzes", *Journal of Archaeological Science: Reports* 39, (2021c) 103163.

O. Oudbashi, M. Mishmastnehi, "Archaeo-metallurgy of copper in the Middle Elamite period of southwestern Iran: Analytical investigation of various parts of the copper production in Haft Tappeh", *Journal of Archaeological Science: Reports* 30 (2020), 102216.

O. Oudbashi, D., Agha-Aligol, M. Mishmastnehi, V. Barnoos, "The Elamite metalworkers: Multianalytical study on copper objects and ingots from second millennium BC of Southwestern Iran", *Archaeological and Anthropological Sciences* 11, (2019), 2059-2072.

O. Oudbashi, A. Hasanpour, "Bronze alloy production during the Iron Age of Luristan: a multianalytical study on recently discovered bronze objects", *Archaeological and Anthropological Sciences* 10 (2018), 1443-1458.

O. Oudbashi, M. Hessari, "Iron Age tin bronze metallurgy at Marlik, Northern Iran: an analytical investigation", *Archaeological and Anthropological Sciences* 9 (2017), 233-249

O. Oudbashi, R. Naseri, M. Malekzadeh, "Technical Studies on the Bronze Age Metal Artefacts from the Graveyard of Deh Dumen, South-Western Iran (Third Millennium BC)", *Archaeometry* 58, (2016), 947-965.

O. Oudbashi, S.M. Emami, P. Davami, "Bronze in Archaeology: A Review of the Archaeometallurgy of Bronze in Ancient Iran", in: L. Collini (ed.), *Copper Alloys-Early Applications and Current Performance-Enhancing Processes*, InTech Open Access Publication, Rijeka, 2012, 153-178.

V.C. Pigott, "On the Importance of Iran in the Study of Prehistoric Copper-based Metallurgy", in: Th. Stöllner, R. Slotta, A. Vatandoust (eds), *Persia's Ancient Splendour, Mining, Handicraft and Archaeology*, Deutsches Bergbau- Museum, Bochum, 2004, 28-43.

V.C. Pigott, H.C. Rogers, S.K. Nash, "Archaeometallurgical Investigations at Malyan: The Evidence for Tin-Bronze in the Kaftari Phase", in: N.F. Miller, K. Abdi (eds), *Yeki bud, yeki nabud: Essays on the Archaeology of Iran in Honor of William M. Sumner*, Monograph 48. Cotsen Institute of Archaeology, University of California, Los Angeles, 2003a, 161-175.

V.C. Pigott, H.C. Rogers, S.K. Nash, "Archaeometallurgical Investigations at Malyan: Banesh Period Finds from ABC and TUV", in W.M. Sumner (ed.), *Malyan Excavation Reports. Vol. III: Early Urban Life in the Land of Anshan: Excavations at Tal-e Malyan in the Highlands of Iran,* The University of Pennsylvania Museum of Archaeology and Anthropology, Philadelphia, 2003b, 94-102.

E.F. Schmidt, *Excavations at Tepe Hissar: Damghan*, University Museum, Philadelphia, 1937.

C.P. Thornton, "The Emergence of Complex Metallurgy on the Iranian Plateau: Escaping the Levantine Paradigm", *Journal of World Prehistory* 22 (2009a), 301-327.

C.P. Thornton, *The Chalcolithic and Early Bronze Age Metallurgy of Tepe Hissar, Northeast Iran: A Challenge to the Levantine Paradigm*, Ph.D. thesis, Department of Anthropology, University of Pennsylvania, 2009b.

C.P. Thornton, C.C. Lamberg-Karlovsky, "A new look at the prehistoric metallurgy of southeastern Iran", *Iran* 42 (2004), 61-76.

C.P. Thornton, C.C. Lamberg-Karlovsky, M. Liezers, M.M. Young, "On Pins and Needles: Tracing the Evolution of Copper-based Alloying at Tepe Yahya, Iran, via ICP-MS Analysis of Common-Place Items", *Journal of Archaeological Science* 29 (2002), 1451-1460.

R.F. Tylecote, *A History of Metallurgy*, 2nd ed. London: The Institute of Metals, 1992.

4. A PERIPHERY? THE ROLE OF NORTHERN GREECE IN THE DEVELOPMENT OF METALLURGY IN THE BRONZE AGE

Zofia Anna Stos-Gale

Abstract

Many publications in the last 30 years concerning lead isotope analysis of Cycladic, Minoan and Mycenaean metals have supported the hypothesis that the production of copper, lead and silver artefacts in the Aegean Bronze Age was based mostly on the exploitation of ores from Lavrion, the Cyclades and Cyprus. This paper attempts a reassessment of the available lead isotope data for Late Neolithic – Late Bronze Age metal artefacts from sites in the Aegean in view of recent discoveries of evidence for early metallurgy of copper, silver and lead in Northern Greece. This reassessment is based on comparisons of several thousand lead isotope data of the metals and ores, mainly from the research programme in Oxford (Isotrace Laboratory) in the years 1978-2001, with added data and archaeological information from more recent publications.

Introduction

The development of the lead isotope analyses of ancient metals has provided a unique scientific tool for provenance studies which has had a significant impact on archaeology, providing additional background to archaeological hypotheses of ancient trade networks and economy of different regions of Europe in the Bronze Age.[1]

The application of lead isotope provenance technique in the research of the Bronze Age metallurgy in the Aegean was initiated by Noel Gale in the late 1970s.[2] Lead isotope provenance studies rely on comparisons of the lead isotope ratios and chemistry of archaeological artefacts with the ores from known ore deposits, therefore the focus of our research has been on parallel sampling of metal artefacts from museums and excavation sites, and surveys of ancient lead/silver and copper mines in the Aegean. Research carried out by the University of Oxford Isotrace Laboratory (located initially in the Department of Earth Sciences and later in the Department of Physics) provided unexpected results relating to the main sources of copper, lead and silver for the Cycladic, Minoan and Mycenaean metal smiths. In contrast to the earlier widely held belief that in the Bronze Age the inhabitants of the Aegean were importing copper (and silver) from a wide area,[3] the lead isotope analyses indicated that throughout the 3rd and 2nd millennia BC the main source of lead, copper and silver were the deposits in Lavrion in Attica,[4] while in the Early Bronze Age, copper, silver and lead were also exploited and smelted on the Cycladic islands.[5] The lead isotope analyses have also resolved Harding's discussion of the origin of the oxhide ingots proving that all of the ingots dated to the 14th-12th centuries BC originated from Cyprus.[6]

The results of this research project based in Oxford seem to indicate that the significant ore deposits in northern Greece did not play an important role as metal sources for the southern part of Mainland Greece and the islands from the 3rd to the end of the 2nd millennium BC.

Ore deposits and slag heaps in Northern Greece

In the years 1978-1996 in collaboration with the geologists from the Greek Geological service IGME and with the assistance in surveys from Stavros Papastavrou, Konstantinos Asimenos and Artemis Papastamataki, and many others, we surveyed 32 mineral occurrences, slag heaps and ancient mines in Greece to collect samples for analyses in Oxford. The map on Fig. 1 shows the approximate position of the sampled ore and slag occurrences. This program of surveys allowed us to create a lead isotope reference

1 Artioli *et al.* 2020; Ling *et al.* 2014 and 2019; Stos-Gale 2019; Rohl and Needham 1998; Radivojević *et al.* 2019.

2 Gale and Stos-Gale 1981 and 1982.

3 Harding 1984, 49-55.

4 Gale *et al.* 2008 and 2009; Kayafa *et al.* 2000; Stos-Gale 2000 and 2014; Stos-Gale *et al.* 1999.

5 Gale and Stos-Gale 2008.

6 Stos-Gale *et al.* 1997.

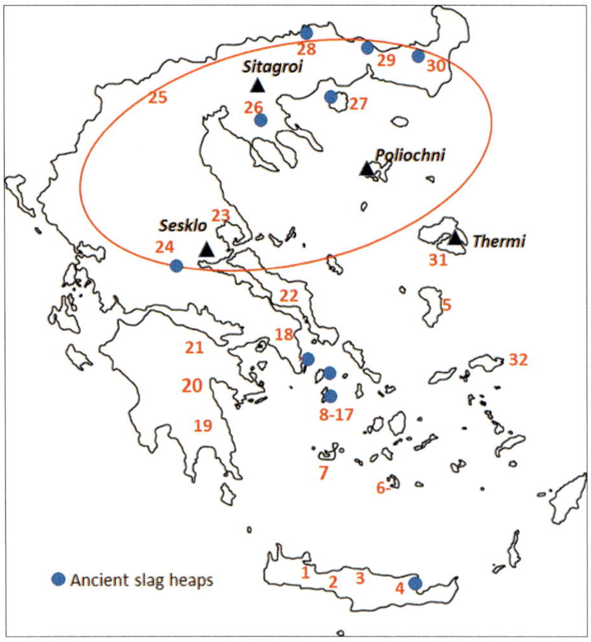

Fig. 1. Map of the ore deposits, small occurrences and ancient slag heaps in Greece surveyed and sampled for lead isotope analyses by the members of the Isotrace Laboratory, Oxford, in collaboration with the geologists from IGME: 1. Sklavopoulo, Kambanos (Fe,Cu): 2. Chrysostomos (Cu), Meskla Miamou, Lebena, Lasaia (Pb, Cu); 3. Ano Varsamonero (Pb) ; 4. Chrysokamino (Cu slag); 5. Chios (Pb); 6.Thera; 7. Melos (Pb); 8-17. Anaphi, Makronissos, Naxos, Poliegos, Syros, Tinos (Pb), Seriphos, Siphnos Kea, Kythnos (Cu, Pb); 18. Lavrion (Cu, Pb, Ag); 19. Molai, Krokees, Apidia (Pb, Cu); 20. Tiros (Pb): 21. Ermioni (Pb): 22. Euboea (Pb): 23. Pelion (Cu,Pb); 24. Othrys Mts. (Cu); 25. Western Macedonia (Cu,Pb), 26. Chalkidiki (Pb,Ag, Cu); 27. Thasos (Pb, Ag, Au, Cu); 28. Pangaeon Mts. (Pb, Cu); 29. Farasinon, Iasmos (Pb); 30. Kirki, Essimi, Pevka (Pb, Cu);31. Lesbos (Pb); 32. Samos. The ellipse indicates the region discussed in this paper.

database for research into the sources of metals used in the Aegean in the Bronze Age. This project funded over the years by the UK Research Councils, the British Academy, Leverhulme Trust and INSTAP resulted in a series of publications and, most importantly, an open access database OXALID (http://oxalid.arch.ox.ac.uk) which is widely used by researchers.[7] In the first years of this survey we collected samples from the major copper, lead and silver occurrences in northern part of Greece marked on Fig. 1 with an ellipse. In this region the evidence of smelting of copper and lead in ancient times is clearly visible by the presence of large slag heaps in the Othrys mountains, in Chalkidiki, on the island of Thasos and in the Rhodope mountains in the region of Xanthi. However, to my knowledge none of these visible slag heaps have been dated to the Bronze Age.

Unfortunately, the published lead isotope data for the northern Greek ore occurrences listed in Table 2 is not very extensive, counting not many more then 200 analyses for all of them, compared with over 500 data just for the ores from Cyprus and over 160 for the region of Lavrion in Attica (OXALID). However, even this limited number of lead isotope ratios should allow the possibility of testing them as possible sources of the Bronze Age artefacts.

Region	Number of available LI data
Pelion	17
Othrys Mountains	33
Western Macedonia	14
Chalkidiki	33
Pangaeon Mountains	25
Eastern Rhodope	39
Thasos	31

Table 1. Number of analysed samples of ores from the Cu/Pb/Ag ore deposits in Northern Greece.

In the recent years there have been several interesting publications revealing Neolithic and Early Bronze Age metallurgical activities on the island of Thasos[8] and in Kastro Palaia near Volos.[9] These publications do not include lead isotope analyses of metallurgical materials and ores, but reasonably assume that the local sources were used. Thus they do not provide information of the specific isotope characteristics of these finds that could be used for comparing their "fingerprint" with metals from other regions.

An extensive review of ore deposits in Greece that could have been used as sources of silver in the 2nd half of the first millennium BC for production of

7 Stos-Gale and Gale 2009.

8 Nerantzis *et al.* 2013, 2016, 2017; Bassiakos *et al.* 2019.

9 Asderaki-Tzoumerkioti *et al.* 2018.

coins, has been published recently.[10] This includes new lead isotope analyses of the ores from Pangeon. Chalkidiki, Pelion and Thasos.

In summary, the ore deposits in northern Greece that could have been sources of silver and copper in antiquity are:

- **Pelion** (Cu, Pb)
  Small occurrences, some ancient mining and LBA copper metallurgy in Kastro Palaia, but no proof of Bronze Age copper exploitation.[11]
- **Othrys** Mountains (Cu)
  Medium size copper deposit, no proof of Bronze Age copper smelting.
- **Chalkidiki** (Pb, Ag, Cu)
  Exploited in 1st millennium BC, Roman and later periods for copper, silver and lead – no proof of BA exploitation of copper or silver.[12]
- **Pangaeon Mts.** (Au-Ag, Cu, Pb)
  1st millennium BC silver exploitation.[13]
- **Palea Kavala**
  (opposite Thasos) and Angistron (NW of Drama) (Ag) – no dating for ancient mining but indicated as possible silver source for the Archaic Greek coinage.[14]
- **Eastern Rhodope**
  Farasinon, Iasmos (Pb); Kirki, Essimi, Xanthi (Pb, Cu) – potential copper source for BA, many slag heaps, but no archaeometallurgical survey.
- **Thasos** (Pb, Ag, Au, Cu)
  Archaeometallurgical evidence of EBA copper smelting and silver extraction. Ag extraction 4th millennium BC,[15] Cu smelting 3rd-2nd millennium BC.[16]

It seems that on current evidence the only copper, lead and silver ore occurrences in northern Greece that have archaeometallurgical evidence of metal extraction earlier than the 1st millennium BC are on the island of Thasos. So, the question is, can we find metals from Thasos among metal artefacts from the Bronze Age sites in Greece?

10 Vaxevanopoulos *et al.* 2022.

11 Asderaki-Tzoumerkiotl *et al.* 2018.

12 Wagner *et al.* 1986.

13 Stos-Gale and Davis 2020; Vaxevanopoulos *et al.* 2022.

14 Vaxevanopoulos *et al.* 2022.

15 Bassiakos *et al.* 2019.

16 Nerantzis and Papadopoulos 2013 and 2016; Nerantzis *et al.* 2017; Bassiakos *et al.* 2019.

Lead isotope analyses of the northern Greek ore deposits and Bronze Age metals from the Aegean

Comparisons of the available lead isotope data indicates that the copper and lead ores from Thasos could have been formed at the same geological events as the ores in the north-west Turkey and Chalkidiki, creating a very similar range of lead isotope ratios. The plots of the lead isotope ratios to ^{204}Pb show that the currently available data for these three regions does not overlap completely (Fig. 2). The ores from Chalkidiki are mostly the occurrences of galena and they are similar only to part of the range of the ores from Thasos. The same can be said for the ores from the region of north-west Turkey, which according to Wagner, Begemann and their colleagues could have been exploited in the Bronze Age.[17] However, they also admit that these ores have never been a major source of copper in the Bronze Age. The lead isotope ratios of galena from Pelion also overlap partly with the ores from Thasos, but there is no evidence of Bronze Age copper smelting from these ores, as discussed above.

The most significant partial overlap of the ores from Thasos is with the LI characteristics of the slags from the 3rd millennium BC (EC II) copper smelting site of Skouries on the Cycladic island of Kythnos.[18] The slag heap of Skouries is of a considerable size and copper smelting there has been securely dated to the Early Cycladic period. This activity therefore was contemporary with the copper smelting on Thasos as described by Nerantzis, Papadopoulos, Bassiakos and their colleagues.

In the 24 years of our research activity in Oxford we analysed over 2000 metal artefacts from 112 archaeological sites in Greece dated to all periods of the Bronze Age (Table 2). Therefore, together with the new lead isotope and archaeological information, these data offer an opportunity to explore the possibility of some of them originating from ore deposits located in northern Greece, and from Thasos in particular.

It seems reasonable to look at first at the comparisons of the lead isotope analyses of metal artefacts from the Bronze Age sites in Thasos with the ores and slags from Thasos. Some 30 years ago we have analysed in the Isotrace Laboratory, University of Oxford, 38 metals from the excavations on this

17 Begemann *et al.* 2003.

18 Stos-Gale 1989 and 1998 and Oxalid.

2

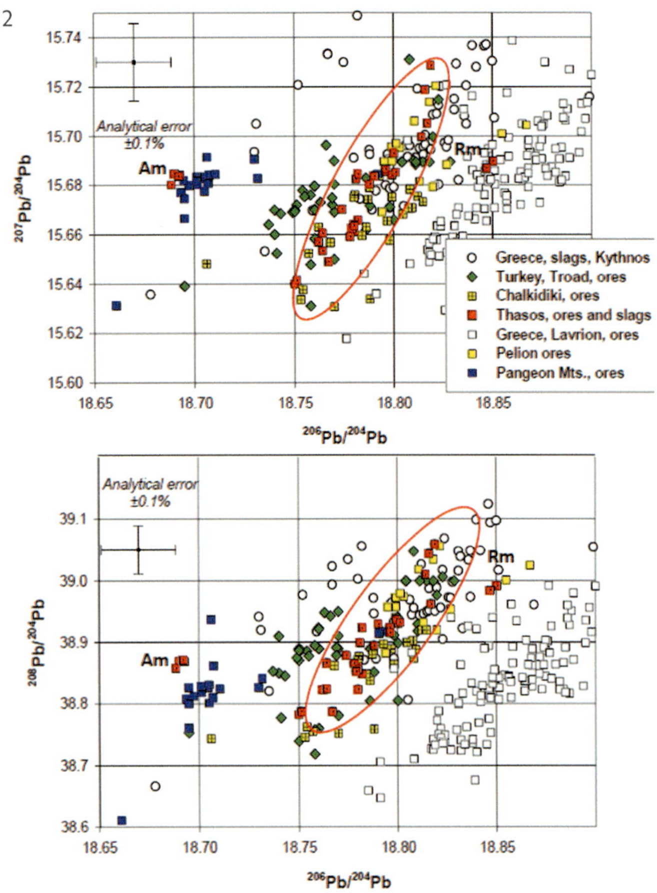

3

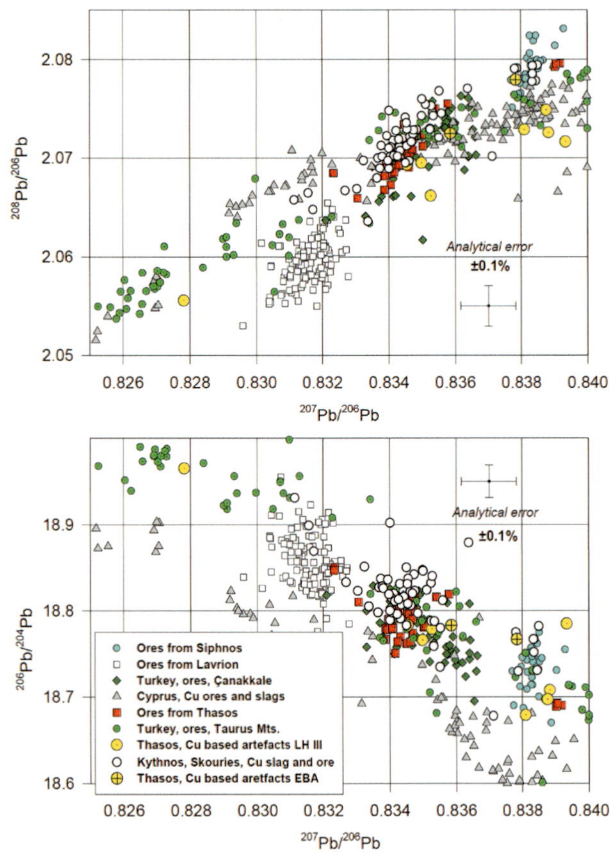

Fig. 2. Plot of lead isotope ratios of the ores from Northern Greece and n-w Turkey, compared with the data for the Cu-Ag-Pb ores from the mines of Lavrion in Attica and copper slags dated to the 3rd millennium BC from the site of Skouries on the Cycladic island of Kythnos. 'Am' (Acropolis mine) and 'Rm' (Rachoni mine) indicate the data for the two mines on Thasos that have different geochronology than most of the copper and lead ores on the island (Am is older, Rm younger than the majority of ores). The ores from Othrys are omitted, because their lead isotope ratios do not overlap with any of the others considered here.

Fig. 3. Comparative plot of the lead isotope ratios of the relevant ores from the Aegean and of the few lead isotope data for the metal artefacts found on Thasos that are dated to the Bronze Age.

Artefacts analysed	No. of sites	LN	EBA	MBA	LBA	Later	Cu	Pb	Ag	Total
Mainland	37	11	47	61	552	150	377	395	49	821
Crete	39	0	56	88	539	7	569	113	8	690
Islands	36	12	191	69	165	60	312	176	10	498
Total	112	22	294	218	1256	217	1258	684	67	2009

Table 2. Summary of lead isotope analyses of metal artefacts from the Aegean dated to the Bronze Age (Isotrace Laboratory, University of Oxford, data on the OXALID).

island led by Ch. Koukouli-Chrysanthaki.[19] This body of material included two Early Bronze Age copper-based fragments, 11 Proto-Geometric, 18 late LH IIIc (1100-1050BC) and 7 dated to the later 1st millennium BC bronzes. The lead isotope data for copper-based metals from Thasos dated to the Bronze Age shows that the majority of them cannot have originated from the ores found on Thasos (Fig. 3). Some of them show close isotopic affinity to the ores from the Taurus Mountains in southern Turkey and

19 Stos-Gale and Gale 1992; the data also on the Oxalid.

4

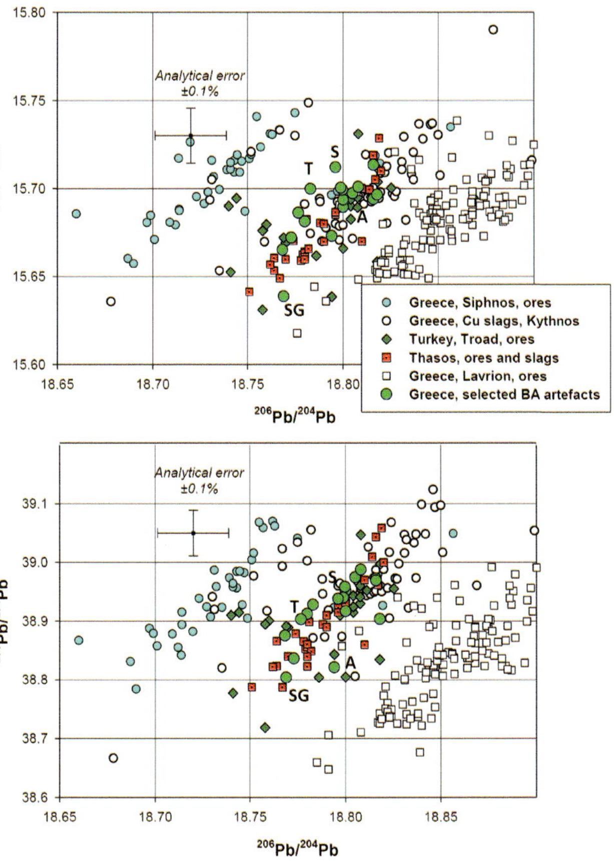

5

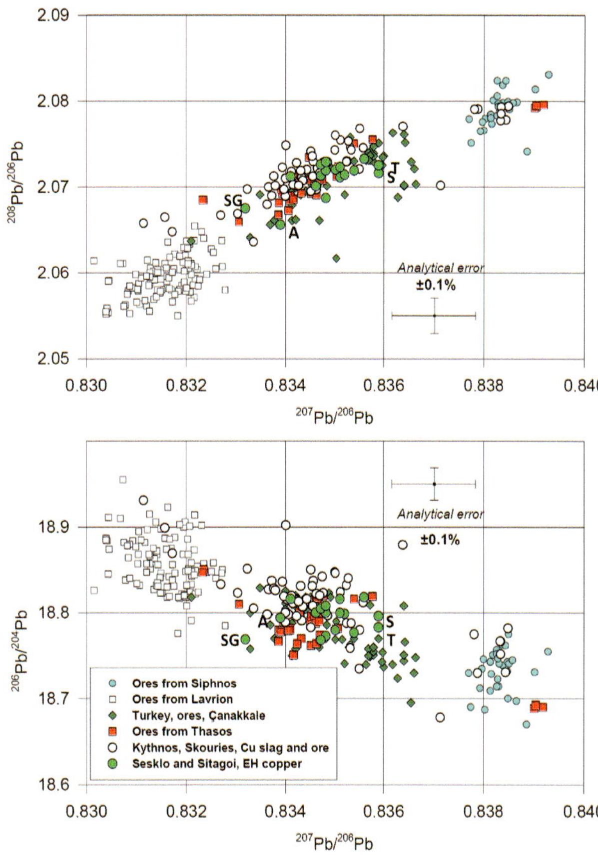

Fig. 4. Comparison of the non-radiogenic (^{204}Pb) lead isotope ratios for the copper and lead ores from the deposits in the vicinity of Thasos that have very similar lead isotope ratios and the copper slags from Kythnos with the data for the BA metals from the Aegean listed in this paper as isotopically consistent with the Thasian ores. The four metals that are not fully consistent with the origin from Thasos are marked on this plot: T- Thasos 719, A- Amorgos AE 236, SG – Mycenae 11310 beta, S- Sitagroi 3005/ROc 30.

Fig. 5. The same data as on Fig. 6 plotted in respect to the radiogenic (^{206}Pb) lead isotope ratios confirms that the 4 metals listed as A, S, T and SG are consistent within the analytical error with the current data for the copper ores from Thasos.

Cyprus. Only two of them are consistent within the analytical error with the lead isotope ratios of the Thasian copper ores: one is the No. 719 fragment dated to EBA, and the other, a bead K177 from the settlement of Kastri dated to late LHIII, excavated by Koukouli-Chrysanthaki, but both of them could also have originated from ores in the north-west Turkey, or in the case of the EBA fragment, from copper smelted on the Cycladic Island of Kythnos. The knives dated to the end of the 2nd millennium are not made of Thasian copper. The interesting aspect of the analyses of these metals is a total lack of metal consistent with the origin from Lavrion. Throughout the Bronze Age copper from Lavrion was omnipresent amongst the Aegean metals[20] and the lack of this source amongst the Group from Kastri on Thasos might indicate that the exploitation of copper from the mines in Lavrion ceased by 1100 BC (Fig. 3).

However, the lead isotope data accumulated during our research in the Isotrace laboratory revealed that amongst the artefacts from other regions of the Aegean there are some that could have originated from the copper and silver ores smelted on Thasos in the Bronze Age (Figs 4 and 5). Amongst the BA artefacts analysed in the Isotrace Laboratory there are 11 copper-based and 2 silver artefacts that have lead isotope ratios fully consistent with the ores found on Thasos. Two silver artefacts are dated to the Early Bronze Age and they are: a bowl from the island of Amorgos in the collection of the Ashmolean Museum, Oxford,[21] and a silver ring from Manika

20 Stos-Gale 2000.

21 Gale and Stos-Gale 1981, p. 211.

on Euboea No. 5515.[22] Out of the 11 copper-based EBA artefacts that have lead isotope ratios consistent with the Thasian ores, 9 are dated to the Early Bronze Age:

- a hook from Emporio on Chios (EMP 11) (unpublished)
- a hook from Kastri on Syros (16176)
- a chisel and a dagger from Kos (M3 and M5) (unpublished)
- an axe from the so called "Kythnos Hoard" found on Naxos (1969/12.31.5)
- a chisel from Poliochni on Lemnos (7209),
- a fragment of copper from Sitagroi (2057/PO/D 25)
- a punch from Thermi on Lesbos (LM4891)
- a chisel (10/L7.202) from EHII-III Lerna in Argolid (Kayafa *et al.* 2000)

There are on our database another three EBA copper-based artefacts that have lead isotope ratios within the analytical error of the ores from Thasos: they are a flat axe from Amorgos from the collection of the Ashmolean Museum (AE236), a fragment of copper from Sitagroi (3005/Roc 30) and the implement from Thasos No. 719, mentioned earlier.

Two LBA copper-based artefacts also could have originated from the Thasian ores. A rivet from a vessel in Nichoria, Messenia (N60; KM 975;1795)[23] dated to the LHI-II is fully consistent with the ores from the Koumaria mine on Thasos. Of great interest here is the copper body of a hydria from the Shaft Graves in Mycenae that has lead isotope ratios within the analytical error of the ores from the mine of Marlou, because this metal is not consistent with any other copper ores and there is a hypothesis that some of the silver vessels in the Shaft Graves have been imported from the north, from Transylvania.[24]

Another interesting suggestion is that a chisel from Dolnoslav, an EBA site in Bulgaria[25] has lead isotope ratios closer to the ores from Thasos, than from Kythnos, as it has been suggested before.[26]

The lead isotope ratios of all artefacts listed above are consistent with the lead isotope ratios of the Thasian mines of Marlou, Koumaria, Makrirachi and Vouves.

22 Stos-Gale *et al.* 1999.

23 Stos-Gale *et al.* 1999.

24 Stos-Gale 2014.

25 Gale *et al.* 2000 and 2003.

26 Stos-Gale and Bajenaru 2020.

All lead isotope data for these artefacts under these numbers can be found on the OXALID.[27]

Conclusions

The recent archaeometallurgical research in northern Greece, and in particular on the island of Thasos, has suggested that perhaps Northern Greek ore deposits were exploited much earlier than the 1st millennium BC. The comparison of the lead isotope characteristics of over 2000 metals from the Aegean dated to the Bronze Age with the available data for the copper, lead and silver ores from Thasos indicate that while this was not a major source of metal at that period for the Aegean metalsmiths, there is a possibility that some copper and silver extracted on Thasos has been circulating in the eastern Mediterranean. The possibility that copper metal on the ECIII site of Kastri that contains metal of a similar origin as metals found in Troy,[28] in the EBA Bulgarian site of Dolnoslav and in LHI Mycenae might indicate that the copper and silver smelted on Thasos have been traded towards the north mainland, rather than to the south. However, these results have to be treated with caution because the lead isotope characteristics of ores from Thasos are not unique, and at present there is no additional information about the chemistry of trace elements that could be specific for the Thasian ores.

The geochemical characterisation of Thasian ores is very limited and therefore at present the above statement can only be treated as a hypothesis. There is a need for much more extensive investigation of the ancient mining on this island to estimate the chronology and extent of mining in the Bronze Age. It would be also good to analyse more copper, silver and lead artefacts from the archaeological sites in northern Greece.

Acknowledgements

This widely collaborative research project was initiated by Noel Gale and was the main focus of the last 30 years of his life. He persevered with the refining and redirecting his knowledge and experience as a physicists and geochronologist into application to provenance of archaeological materials despite dismissive criticism and competitive obstacles. The analytical work was made possible by 25 years of funding from the UK Research Councils, the British Academy, the University of Oxford, the Leverhulme

27 http://oxalid.arch.ox.ac.uk

28 Stos-Gale *et al.* 1984.

Trust and INSTAP. Wonderfully friendly support of the numerous geologists and archaeologists, and the British School in Athens enabled the collection of samples for analyses.

Bibliography

G. Artioli, C. Canovaro, P. Nimis and I. Angelini, , "LIA of prehistoric metals in the central Mediterranean area: a review", *Archaeometry* 62, Suppl. 1 (2020) 53-85. DOI**:** 10.1111/arcm.12542

Y. Bassiakos, N. Nerantzis and S. Papadopoulos, "Late Neothic/Early Bronze Age metallurgical practices at Limenaria, Thasos: evidence for silver and copper production", *Archaeological and Anthropological Sciences* 11/6 (2019), 2743-2757. https://doi.org/10.1007/s12520-018-0689-4.

F. Begemann, S. Schmitt-Strecker and E. Pernicka, "On the compositions and provenance of metal finds from Besiktepe (Troia)", in: G.A. Wagner, E. Pernicka, H.-P. Uerpmann (eds), *Troia and the Troad*, Berlin: Springer, 2003, 173-201.

N.H. Gale and Z.A. Stos-Gale, "Lead and silver in the ancient Aegean", *Scientific American* Vol. 244, No. 6 (1981) 176-192.

N.H. Gale and Z.A. Stos-Gale, "Cycladic lead and silver metallurgy", *Annual of the British School at Athens*, 76 (1981) 169-224.

N.H. Gale and Z.A. Stos-Gale, Bronze Age Copper Sources in the Mediterranean: A new approach. *Science,* Vol. 216 (1982) 11-19.

N.H. Gale, Z.A. Stos-Gale, A. Radountcheva, I.Ivanov, P. Lilov, T. Todorov and I. Panayotov, *Early Metallurgy in Bulgaria. Godishnik Nov Bulgarski Universitet*, IV-V: 102-68, 2000.

N.H. Gale, Z.A. Stos-Gale, A. Radountcheva, I. Ivanov, P. Lilov, T. Todorov and I. Panayotov, **"**Early Metallurgy in Bulgaria**"**, in: P. Craddock and J. Lang (eds), *Mining and Metal Production Through the Ages*. London: British Museum Press, 2003, 122-173.

N.H. Gale, M. Kayafa and Z.A. Stos-Gale, "Early Helladic Metallurgy at Raphina, Attica, and the Role of Lavrion", in: I. Tzachili (ed.), *Aegean Metallurgy in the Bronze Age*, University of Crete, Rethymnon, Athens, Ta Pragmata Publications, 2008, 87-104.

N.H. Gale, Z.a. Stos-Gale, "Changing Patterns in Prehistoric Cycladic Metallurgy", In: N. Brodie, J. Doole, G. Gavalas and C. Renfrew (eds), *Horizon: A Colloquium on the Prehistory of the Cyclades*, Cambridge: McDonald Institute for Archaeological Research, 2008, 387-408.

N.H. Gale, M. Kayafa and Z.A. Stos-Gale, "Further evidence for Bronze Age production of copper from ores in the Lavrion ore district, Attica, Greece", in: A. Giumlia-Mair, P. Craddock, A. Hauptmann, J. Bayley, M. Cavallini, G. Garagnani, B. Gilmour, S. La Niece, W. Nicodemi and T. Rehren (eds), *Proceedings of the 2nd International Conference: Archaeometallurgy in Europe, 2007*, Milan, 2009, 158-76.

A. Harding, *The Mycenaeans and Europe*. Emerald Group Publishing. Bingley, UK, 1984.

M. Kayafa, Z.A. Stos-Gale and N.H. Gale, "The circulation of copper in the Early Bronze Age in Mainland Greece: the lead isotope evidence from Lerna, Lithares and Tsoungiza", in: C. Pare (ed.), *Metals make the world go round: supply and circulation of metals in Bronze Age Europe*, Oxbow Books, Oxford, 2000, 39-55.

J. Ling, Z.a. Stos-Gale, L. Grandin, K. Billström, E. Hjarthner-Holdar, P.O. Persson, "Moving metals II: provenancing Scandinavian Bronze Age artefacts by lead isotope and elemental analyses", *Journal of Archaeological Science* 41 (2014) 106-132.

J. Ling, E. Hjarthner-Holdar, L. Grandin, Z.A. Stos-Gale, K. Kristiansen, A.L. Melheim, G. Artioli, I. Angelini, R. Krause, C. Canovaro, "Moving Metals IV: Swords, metal sources and trade networks in Bronze Age Europe", *Journal of Archaeological Science: Reports* 11: (2019) 4829-4846.

N. Nerantzis and S. Papadopoulos, "Reassessment and new data on the diachronic relationship of Thasos Island with its indigenous metal resources: a review", *Archaeological and Anthropological Sciences* 5/3, 2013, 183-196.

N. Nerantzis, Y. Bassiakos and S. Papadopoulos, "Copper metallurgy of the Early Bronze Age in Thasos, north Aegean", *Journal of Archaeological Science: Reports* 7, 2016, 574-580.

N. Nerantzis, Y. Bassiakos, M. Georgakopoulou, E. Filippaki and G. Mastrotheodoros, "Copper production in Early Bronze Age Thasos: new finds and experimental simulation in the context of contemporaneous Aegean metallurgical practices", in: I. Montero-Ruiz and A. Perea (eds), *Archeometallurgy in Europe IV*, Madrid, 2017, 115-125.

E. Pernicka and G.A. Wagner, "Thasos als Rohstoffquelle für Bunt- und Edelmetalle im Altertum", in: G.A. Wagner and G. Weisgerber (eds), *Antike Edel- und Buntmetallgewinnung auf Thasos.* Der Anschnitt 6. Bochum: Deutsches Bergbau Museum, 1988, 224-231.

M. Radivojević, B.W. Roberts, E. Pernicka, Z.A. Stos-Gale, M. Martinón-Torres, T. Rehren, P. Bray, D. Brandherm, J. Ling, J. Mei, H. Vandkilde, K. Kristiansen, S.J. Shennan, C. Broodbank, "The provenance, use, and circulation of metals in the European Bronze Age: the state of debate", *Journal of Archaeological Research* 27, 2019, 131-185.

B. Rohl and S. Needham, *The circulation of metal in the British Bronze Age: the application of lead isotope analysis.* British Museum, London, British Museum Occasional Paper 102, 1998.

Z.A. Stos-Gale, N.H. Gale and G.R. Gilmore, "Early Bronze Age Trojan metal sources and Anatolians in the Cyclades", *Oxford Journal of Archaeology* Vol. 3/3, 1984, 23-44.

Z.A. Stos-Gale, "Cycladic copper metallurgy", in: A. Hauptmann, E. Pernicka and G.A. Wagner (eds), *Old World Archaeometallurgy, Der Anschnitt* 7, Bochum: Bergbau Museum, 1989, 279-291.

Z.A. Stos-Gale and N.H. GALE, "Sources of copper used on Thasos in Late Bronze and Early Iron Age", in: Ch. Koukouli-Chrysanthaki (ed.), Πρωτοϊστορική Θάσος, Τα νεκροταφεία του οικισμού Καστρί, Αθήνα: Δημοσιεύματα του Αρχαιολογικού Δελτίου, Αρ. 45, 781-793.

Z.A. Stos-Gale, **"**The role of Kythnos and other Cycladic islands in the origins of Early Minoan metallurgy**"**, in: L. Mendoni and A. Mazarakis-Ainian (eds), *Kea-Kythnos: History and Archaeology, Meletimata* 27, Athens, 1998, 717-735.

Z.A. Stos-Gale, G. Maliotis, N.H. Gale and N. Annetts, "Lead isotope characteristics of the Cyprus copper ore deposits applied to provenance studies of copper oxhide ingots", *Archaeometry* 39 (1), 1997, 83-124.

Z.A. Stos-Gale, M. Kayafa and N.H. Gale, "The origin of metals from the Bronze Age site of Nichoria", *Opuscula Atheniensia,* 24. Lund. 1999, 99-120.

Z.A. Stos-Gale, A. Sampson and E. Mangou, "Analyses of metal artefacts from the Early Helladic cemetery of Manika on Euboea", *Aegean Archaeology,* Vol 3. 1996. Art and Archaeology, Warsaw, 1999, 49-62.

Z.A. Stos-Gale, "Trade in Metals in the Bronze Age Mediterranean: an overview of lead isotope data for provenance studies", in: C. Pare (ed.), *Metals make the world go round: supply and circulation of metals in Bronze Age Europe*, Oxford: Oxbow Books, 2000, 56-69.

Z.A. Stos-Gale and N.H. Gale, "Metal provenancing using isotopes and the Oxford archaeological lead isotope database (OXALID)", *Archaeological and Anthropological Sciences*, Volume 1, Number 3, 2009, 195-213.

Z.A. Stos-Gale, "Silver vessels in the Mycenaean Shaft Graves and their origin in the context of the metal supply in the Bronze Age Aegean", in: H. Meller, R. Risch and E. Pernicka (eds), *Metalle der Macht_ Frühes Gold und Silber.* 6. Mitteldeutscher Archäologentag vom 17. Bis 19 Oktober 2013 in Halle (Saale). Tagungen des Landesmuseums für Vorgeschichte Halle. Band 11/1, 2014, 183-208.

Z.A. Stos-Gale, "Lead isotopes and the origin of bronze artefacts" in: J. Sobieraj (ed.), *The origins of the Bronze Age in Warmia and Masuria revealed by scientific analyses.* Muzeum Warmii i Mazur, Olsztyn, 2019, 83-118.

Z.A. Stos-Gale and G. Davis, "The Minting/ Mining Nexus: New Understandings of Archaic Greek Silver Coinage from Lead Isotope Analysis", in: K.A. Sheedy and G. Davis (eds), *Metallurgy in Numismatics 6, Mines, Metals, and Money*, Ancient World Studies in Science, Archaeology and History. Australian Centre for Ancient Numismatic Studies, Macquarie University, The Royal Numismatic Society Special Publication No.56. London, 2020, 87-100.

Z.A. Stos-Gale and R. Bajenaru, "The Aegean and the Black Sea connecting southeast Europe and Anatolia during the Early bronze Age: evidence from metal finds in Bulgaria, Greece and western Turkey" in: J. Maran, R. Bajenaru, S.-C. Ailincai, A.d. Popescu and S. Hansen (eds), *Objects, ideas and travellers. Contacts between the Balkans, the Aegean and Western Anatolia during the Bronze and Early Iron Age,* Universitätsforschungen zur Prähistorischen Archäologie, Institut für Ur- und Frühgeschichte der Universität Heidelberg, Band 350. Verlag Rudolf Habelt Gmbh, Bonn, 2020, 265-285.

M. Vaxevanopoulos, J. Blicher-Toft, G. Davis and F. Albarède, "New findings of ancient Greek silver sources", *Journal of Archaeological Science.* 137, 2022, 105474.

G.A. Wagner, E. Pernicka, M. Vavelidis, I. Baranyi and Y. Bassiakos, *Archäeometallurgische Untersuchungen auf Chalkidiki.* Der Anschnitt 38. Bochum: Deutsches Bergbau Museum, 1986.

G.A. Wagner and G. Weisgerber, *Antike Edel- und Buntmetallgewinnung auf Thasos, Der Anschnitt,* Beiheft 6, Bochum: Bergbau-Museum, 1988.

5. CHANGING PATTERNS IN THE EXPLOITATION, USE AND SIGNIFICANCE OF GOLD IN THE NORTHERN AEGEAN (1200-600 BCE)*

Elon D. Heymans and Samuel Verdan

Introduction

Since the 1980s, archaeological, archaeometallurgical and geological research has made a huge impact on our knowledge of mineral resources, such as metal sources, in the ancient Aegean.[1] However, scholarship on the exploitation of such resources, and the history of mining in general, has often focused on technological or economic perspectives, upholding an artificial distinction between human actors and a passive natural world, whereby metal resources are implicitly perceived as waiting in the earth's crust until the technological and/or economic conditions are ripe for exploitation and the metal can be put to a set use. By contrast, if we turn our perspective around and emphasize the dynamic relation between human populations and their natural environment in the ancient world, we have a lot to gain in understanding how metal resources helped shape new values and how changing patterns of exploitation and use, as well as the human interactions involved, can be tied to changes in history and culture.[2] To this end, we will focus on the role of natural gold sources and their exploitation in the changing societies of the northern Aegean between the end of the Late Bronze Age and the archaic period.

The northern Aegean is rich in natural gold sources. Numerous primary and placer gold deposits were known in antiquity, such as the various gold-bearing rivers and streams around the Thermaic gulf and western Chalkidiki.[3] Prominent among these are the Haliakmon, Axios and Gallikos rivers, passing through the central Macedonian plain before flowing into the Thermaic gulf (Fig. 1). While gold sources were long known in the area, presumably going back to the Neolithic,[4] it is not in and of itself evident that they were continuously exploited or that gold was even used and valued in a context and on a scale that reflect long-held perceptions of its role as a primary commodity of value and status marker. Rather than seeing gold sources as a static presence in the landscape, we would do well to keep in mind that gold might not have always been easily available for exploitation and that exploitation can also lead to exhaustion. Moreover, it is important to state that human interest in gold mining is neither natural nor self-evident, but emerged (and could subside) under specific circumstances. This is, among other reasons, due to the fact that conceptions of value and the related demand for certain commodities need not be stable or follow a linear evolution.

Nonetheless, limited interest in gold, especially if readily available, should appear counter-intuitive to the modern observer. Gold is the material of which wealth is made and its aesthetic qualities and durability have long insured its use as a status marker. If gold could be collected from riverbeds, would it not be considered advantageous to do so? This brings to mind what anthropologist Raymond Firth wrote about the bonito hook, which was used as a so-called primitive currency among the inhabitants of Polynesian Tikopia:

"I have always thought it remarkable that the Tikopia do not make more bonito-hooks. The question why some sharp individuals do not accumulate a stock for trading purposes and why all men do not put in more labour in the production of them is difficult to answer. It seems that the attitude of the Tikopia

* We want to thank Nerantzis Nerantzis for his efforts in organising the Brussels conference and the other participants for their critical comments.

1 See Vaxevanopoulos *et al.* 2022 for a recent example.

2 Cf. LeCain 2019.

3 Gold sources in the northern Aegean, in particular those within the borders of modern-day Greece, have been explored extensively, especially by professor M. Vavelidis, and the literature on the topic is vast. For references, see e.g. Vavelidis and Andreou 2008; Andreou and Vavelidis 2014.

4 Andreou and Vavelidis 2014, 455-459.

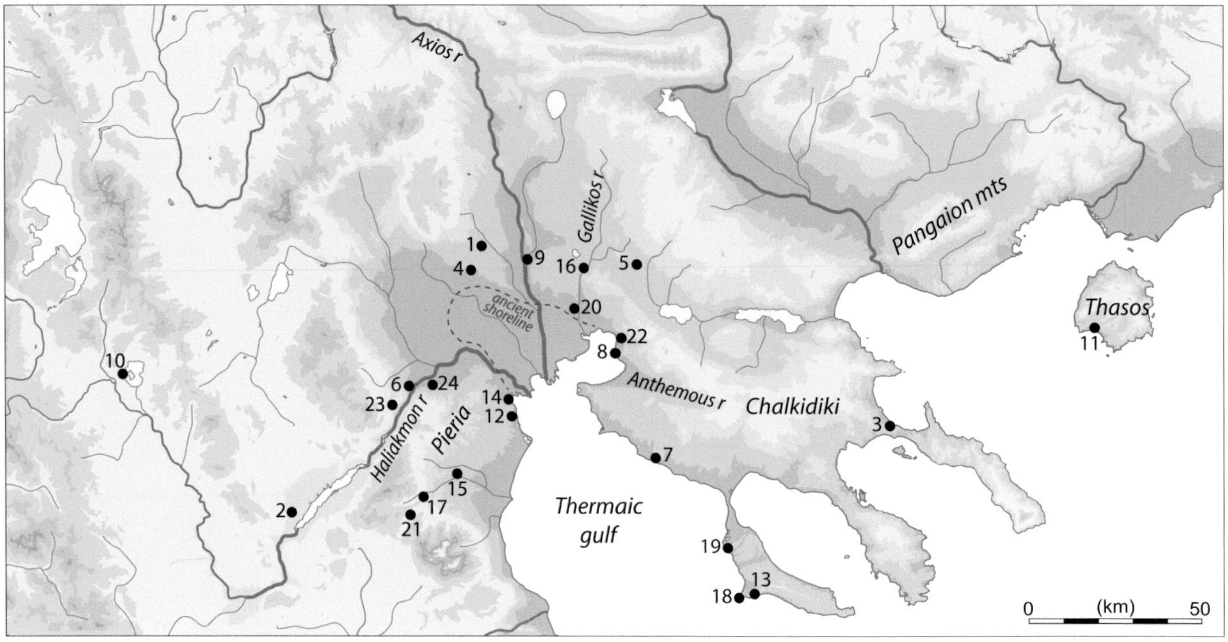

Fig. 1. Map of the northern Aegean with sites mentioned in the text: 1. Agrosykia, 2. Aiani, 3. Akanthos, 4. Archontiko, 5. Assiros, 6. Assomata-Tsardaki, 7. Dikaia?, 8. Karabournaki, 9. Kastanas, 10. Kastoria-Daïlaki, 11. Limenaria, 12. Makrygialos, 13. Mende, 14. Methone, 15. Moschochori, 16. Nea Philadelphia, 17. Petra-Treis Elies, 18. Cape Poseidi, 19. Sane, 20. Sindos, 21. Spathes, 22. Thessaloniki Toumba, 23. Tzamala, 24. Vergina.

in this is governed by other factors than those mentioned. Supply does not necessarily respond to demand because of indifference which is the result of attitudes towards time, labour, and material objects which themselves are part of a wider scheme of social values – indeed, almost of a philosophy of life."[5]

Similarly so, historical examples indicate that communities need not always demonstrate a clear interest in or preference for gold and that specific historical or social circumstances can lead to change in this respect, whether caused by new consumption patterns or the profitability of its exploitation for an emerging external demand. Recent research, for example, has demonstrated a collapse in the consumption of gold objects in the Caucasus – a region rich in gold sources – during the Late Bronze and Iron Age (1500-700). Compared to the Middle Bronze Age, both the quantity and the level of complexity of gold working declined, not as a result of the availability of resources or demographic change but because the conspicuous elite display of gold, associated with the Middle Bronze Age social hierarchy, lost its relevance as a social strategy.[6]

Another striking case to consider are the Akan in western Africa. While gold sources are relatively widespread in sub-Saharan Africa, African societies generally never displayed much interest in it, preferring copper as the primary metal of value.[7] A conspicuous exception are the Akan, in the region of modern Ghana and western Ivory Coast, where gold increasingly featured as a currency since the 17th century, being weighed out as gold dust using a system of cast brass weights.[8] Interestingly though, exploitation of the rich gold sources native to the region only took off in the 15th century as a result of contact with Mande traders from the north and was further stimulated as European colonists were drawn into the gold trade.[9] Increased production of gold, peaking in the 17th century, gave way to new expressions of social and political hierarchy, and resulted in political developments such as the emergence of the Asante empire, which managed to control much of the gold trade.[10] It thus appears that gold only became a primary commodity of value, symbol of wealth and status marker in Akan society *as a result* of the foreign gold trade that had such a transformative effect on the region.

5 Firth 1965, 342.

6 Erb-Satullo 2021.

7 Herbert 1984, spec. 296-303.

8 Garrard 1980, 32, 48, 50.

9 Garrard 1980, 1-33.

10 Garrard 1980, 59-66, 163-166

Likewise, in the northern Aegean, exploitation of gold sources on a large scale was not continuous, nor was gold always used and valued in the same way. In this paper, we wish to show how interactions between local groups, Euboian traders and the local environment in the region played out over a longer time period, shaping the changing attitudes towards gold and its exploitation. We will suggest that interactions revolving around the exploitation of gold played an important role in shaping Euboian involvement in the region, while Euboian demand, in turn, stimulated the appropriation of gold within local regimes of value.

The suggestion that Euboians were attracted to the region due to its gold sources is hardly new: Michalis Tiverios drew attention to the exploitation of gold at Sindos over twenty years ago.[11] However, expanding our approach to encompass a broader timeframe and region will enable us to better contextualize the limited and often indirect evidence. This is relevant because evidence for the exploitation and use of gold is hardly ever straightforward. The processing of gold, especially from placer sources, leaves few traces and is difficult to identify in the archaeological record. Gold *itself* circulated only in small quantities and people were careful to recycle any of it, so unless deliberately deposited, we are not likely to find it in the first place. All the more extraordinary is the Late Geometric gold hoard from Euboian Eretria, the single largest gold find from the Iron Age and Archaic period. Without this chance find, our attention would probably not be focused on this issue in the first place.

Our ongoing work on the Eretria hoard provides valuable insights that enable us to contextualize the changing role of gold in the Aegean, and Euboian involvement in its exploitation. At the same time, through the hoard, our view of Euboian activity in the north can provide a model for other contemporary contexts of gold exploitation, where evidence is even more scrappy.

Northern Aegean gold: local communities and Euboians

During the course of the Iron Age, people from the island of Euboia appear to show an active interest in the northern Aegean, specifically the region of the Thermaic gulf and western Chalkidiki. Whether and when Euboians established permanent and autonomous settlements, and whether or not this should be understood in the framework of colonisation or precolonisation, is part of an ongoing debate.[12] Nonetheless, the presence of Euboian and Euboian-style pottery suggests intensifying contacts with and an increased involvement in the region down to the beginning of the Archaic period.

The early nature of Euboian presence in the Iron Age northern Aegean is not very clear, although it is possible that contacts, established already during the Palatial period, had been maintained and rearranged in the Postpalatial and subsequent periods.[13] The cultic site at Cape Poseidi, west of Mende, contains an ash altar dating to the 12th century with Late Mycenaean pottery around which an apsidal cult building was established in the 10th century and was likely frequented by Euboians.[14] The strategic significance of the site can hardly be over emphasized: the location of the site marked the entrance to the Thermaic gulf as well as sea routes to and from the east, making the site a key stop-over for sailors and merchants.

Gold from the alluvial deposits around the Thermaic gulf – the Haliakmon, Axios, Gallikos and Anthemous rivers – could have attracted traders from the southern Aegean already during the Late Bronze Age, continuing into the Iron Age.[15] Interestingly, however, Macedonian sites from the period offer very limited evidence for the conspicuous consumption of gold objects. Across the region, gold objects are invariably simple and few.[16] Well-excavated settlement sites suggest no particular association between gold and high-status contexts, with scraps found both at the small and lower-level site of Kastanas and at the large centre at Thessaloniki Toumba.[17] From the latter, we also have four small stone crucibles, one of which contained traces of gold working, found in different domestic units. While providing the only evidence we

11 Tiverios 1998, 248-250; 2008, 21; cf. Leone 2012, 232-233.

12 Tiverios 2008, 1-52; Popham 1994; Snodgrass 1994; Papadopoulos 1996; more Recently Kotsonas 2012; Gimatzidis 2020; 2022.

13 Lemos 2020, 803-804.

14 Moschonissioti 1998, 260-267; cf. Papadopoulos 1996; Charalambidou 2017, 98.

15 Vavelidis and Andreou 2008; Saripanidi 2017, 105; see also a current project led by prof. Ernst Pernicka on Mycenaean gold (https://www.gerda-henkel-stiftung.de/mykenisches-gold-untersuchungen-zu-herkunft-verbreitung-und-echtheit?page_id=122228 ; accessed on July 18, 2022). Silver from the shaft graves was previously also argued to have originated from the northern Aegean (see Stos-Gale and Mac-Donald 1991, 273-279; Papadopoulos 2011, 123), but this has now been rejected (Stos-Gale 2014).

16 Andreou 2010, 651-652.

17 Hochstetter 1987, 15-18.

have for gold working in the region for the 12th and 11th century, it confirms to a pattern of small-scale production within the settlement, requiring little specialized knowledge or infrastructure.[18]

In central Macedonia, the inconspicuous role of gold can be understood in the context of social and political norms that regulated communal life. Tall mounds, such as Assiros (Phase 6 and 7) and Thessaloniki Toumba (Phase 4), display an elaborate settlement organisation with a regular layout of streets and housing blocks with storage facilities.[19] There are, however, no central or elite structures that dominate the settlement, and equality and communality appear to have been emphasized over political hierarchy and elite identity.[20] Likewise, as funerary evidence in the region is limited to a few graves inside the settlement of Thessaloniki Toumba, treatment of the dead does not appear to have been a focus of attention or investment.[21] Alternatively, outside of central Macedonia, cemeteries became a prime setting for the expression of social identities, although even in the wealthy burials at Aiani and Spathes in southern Macedonia gold is absent.[22]

Explaining the limited interest in gold as a status marker, Andreou and Vavelidis argue that "the absence of highly complex socio-political formations and strong rules, which prevailed, particularly in communities in central Macedonia, discouraged the expression of intra-community distinctions through the excessive display of wealth in habitation and mortuary contexts."[23] This, together with the relative proximity to and availability of placer gold sources, and the fact that processing or producing gold items required hardly any technological expertise, provided a context for the lower valuation of gold, both as a commodity and as a symbolic marker.[24]

In marked contrast stands Lefkandi on Euboia, where, especially from the MPG onwards, finds testify to the importance of gold as an expression of elite identity. Tombs of wealthy elites in Lefkandi contain many exotic objects that reflect maritime and long-distance contacts.[25] The association with these graves shows that gold should not be seen as merely an expression of wealth per se, but as a form of exotic wealth. Although the many golden objects in Lefkandi could have had diverse origins, from the few cautious suggestions made so far, the northern Aegean is an option worthy to further explore.[26]

Contacts with the northern Aegean are reflected by a variety of finds in Lefkandi, such as transport amphorae, and Irene Lemos even stated that "in every tomb with imports from the East, there are also imports from the North".[27] A good example is the wealthy LPG grave 63 at the Toumba cemetery. Among the many finds were three handmade burnished jugs with cutaway necks, likely imports from the north, and an exquisite golden necklace with quadruple spiral beads.[28] Although parallels are known from Kefalonia and Skyros – a waystation between Euboia and the north – the spectacle-spiral ornament is well known from northern Aegean bronzework, and the necklace could well have been imported, or locally produced, following northern Aegean models.[29] Slightly later, in the course of the 9th century, the same spiral motif appears again, first in two earring pendants in T.13 (SPG II)[30] and next in a more eclectic local adaptation; a golden crescent-shaped pendant from T.38 (SPG II-IIIa) is decorated with a spectacle spiral and triangles of granulation on the crescent-shaped plate.[31] It follows a slightly earlier example from Toumba grave 59 (LPG/SPG I), without spiral but with granulation, forming the earliest evidence

18 Vavelidis and Andreou 2008, 362-363; Andreou and Vavelidis 2014, 460-461.

19 Wardle 1988, 382-384, 387; 1989, 456, Fig. 2; Andreou 2001, 166-168; Andreou and Efkleidou 2011.

20 Andreou 2001; 2010, 649-651; 2020, 915-918.

21 Andreou 2010, 651; Chavela 2018, 163; 2022, 54; Triantafyllou and Andreou 2020, 178-179.

22 Andreou and Vavelidis 2014, 460; in general, see Triantafyllou and Andreou 2020.

23 Andreou and Vavelidis 2014, 461.

24 Cf. Heymans 2021, 229-231.

25 Crielaard 2006; 2012; Leone 2012.

26 R. E. Jones (1980), in his analyses of the gold jewellery, states that the source of the gold could not be identified. For Egypt as a source of the gold, see Descoeudres 2008, 306; for the Near East, see Coldstream 2003, 19; for the northern Aegean, see Tiverios 1998, 249; Leone 2012, 232.

27 Lemos 2001, 217.

28 Popham *et al.* 1989, 120 and 128, Fig. 23; Popham and Lemos 1996, pl. 68-69 and 136j; Martin Pruvot *et al.* 2010, 90-91, cat. 20.

29 Coldstream 1995, 398; Lemos 2002, 132. Northern origins of the spiral motive were already suggested by R. A. Higgins (1980, 221). Skyros: Popham and Lemos 1996, pl. 157f; Lemos 2002, 132. The term, quadruple spiral bead, follows Maxwell-Hyslop (1971, 35-36), although the earlier types she describes from the ancient Near East are produced very differently. Closer to our example are the quadruple and sextuple spiral beads from shaft grave III at Mycenae (Karo 1930, 52, 188, pl. 21.56-59).

30 Popham *et al.* 1979-1980, pl. 173 and 231b.

31 Popham *et al.* 1982, 236 and pl. 30a; Popham and Lemos 1996, pl. 40 and 136a.

for the adoption of eastern techniques.[32] A larger, more exquisite example, adding a central band with interconnected spirals in filigree to the composition of the T.38 pendant, is known from a tomb in Chalcis.[33]

It is during this time, in the 9th and particularly the 8th century, that Euboian presence in the northern Aegean is more firmly established. Most notably at Sindos, the emporion or port of call on the Thermaic gulf, at the mouth of the Gallikos river, Euboians were joining the local population, as suggested by the substantial quantities of Euboian pottery in level 8 and 7 (from the late 9th/early 8th century – SPG IIIb).[34] Michalis Tiverios claimed with regard to this context that "there can be no doubt that the Euboians were attracted here chiefly by the gold in the Gallikos river".[35] Around the same time, Euboians were increasingly formalising their strategic presence in the form of permanent settlements, such as suggested on the basis of the remains around the Vigla hill and the sea-side cemetery at Mende.[36]

On the other side of the gulf, Methone, located on the northern edge of the Pieria, just south of the Haliakmon river delta, provides us with the most important clues for Euboian interest in northern Aegean gold. Following possible earlier contacts, we see clear evidence of an intensification of Euboian involvement in the second half of the 8th century, corresponding to the historical tradition of the foundation of an Eretrian colony there in 733/2, as related by Plutarch.[37]

Gold working at Methone and the Eretria hoard

Methone was an important regional trading centre, as indicated by the ceramic evidence excavated from the socalled *hypogeion*, a deep underground storage facility abandoned and filled in around the beginning of the 7th century – a timeframe closely following the supposed foundation of the colony. Trade amphorae and fine wares from across the Aegean, even including Phoenician transport amphoras, suggest that Methone was frequented by a diverse community of traders and functioned as a central node in a network, connecting the Macedonian hinterland with the Aegean and beyond.[38] Moreover, the *hypogeion* contained important clues regarding the processing of gold at the site.[39] These are coarse ware sherds, identified as melting plates, also known from 8th-century layers in the area of the Apollo sanctuary in Eretria and the 7th-century phase of the Pactolus North area in Sardis. They had been used to melt gold dust from placer deposits into larger globule-shaped ingots, indicated by the tiny globules stuck in the vitrified layer and the rounded impressions left on the sherds as a result of the heavy gold ingots sinking into the molten surface (Fig. 2).[40] Together with a sizable nugget and two cut halves of a hammered ingot (Fig. 3), this evidence suggests that – in addition to being an important trade emporion on the edge of the Thermaic gulf – Methone (as probably Sindos) was used as a place to process the gold collected from the nearby placer sources by remelting the gold dust into larger ingots, more suitable for use and transport.

Fig. 2. Methone: gold melting plate from the *hypogeion* (MEΘ 5331/ME 1483; photo I. Coyle, courtesy of *Ancient Methone Archaeological Project*).

32 Popham *et al.* 1988-1989, 120 and 128, Fig. 25; Popham and Lemos 1996, pl. 66 and 136b.

33 Currently on display in the new archaeological museum of Chalcis: see Simosi 2022, 220.

34 Gimatzidis 2010, esp. 306-308; Gimatzidis 2011, 101; cf. Kotsonas 2012, 249 and 256-257.

35 Tiverios 1998, 248-250; 2008, 21.

36 Moschonissioti 1998; 2017; but see now Gimatzidis 2022.

37 Plut. *Quaest. Graec.* 11 = *Mor.* 293 A-B.

38 Bessios *et al.* 2012, in particular chap. 3 (Kotsonas); for the Phoenician transport amphoras, see Kasseri 2012; Ilieva 2019, 74-75. On Methone more broadly, see Morris and Papadopoulos, 2023.

39 We are very grateful to Manthos Bessios, Sarah Morris and John Papadopoulos for granting us initial access to the Methone finds.

40 Sardis: Ramage and Craddock 2000, 90, 100, 102-117, 126-127, 132-134, 160-161, 207. Eretria: Meeks and Craddock 2013, 271-273, pl. 127-129; Verdan 2013, 148-149, pl. 114. For the Methone melting plates and a detailed description of the production process, see Verdan 2023; Verdan and Heymans 2020.

3

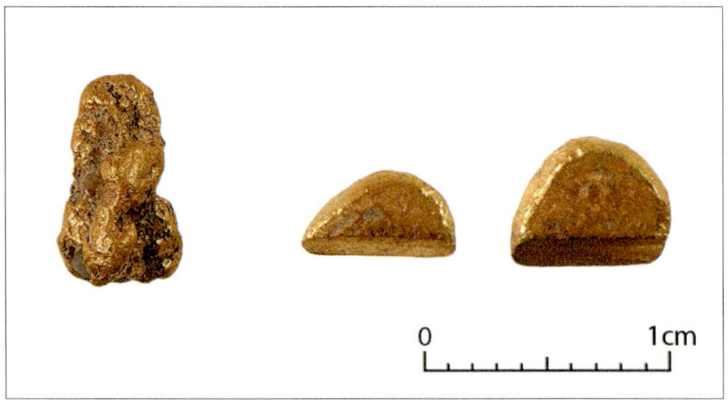

Fig. 3. Methone: nugget and two halves of an ingot (ME 1345 and ME 1346; photo I. Coyle, courtesy of *Ancient Methone Archaeological Project*).

Fig. 4. Eretria: Late Geometric gold hoard, general overview shortly after discovery (photo P. G. Themelis).

Fig. 5. Eretria hoard: left) dendrites on the upper surface of an ingot (Mus. Eret. inv. 14952/no. 5); right) dendrites visible in a cut ingot (Mus. Eret. inv. 14958/no. 300; photo P. Simon).

4

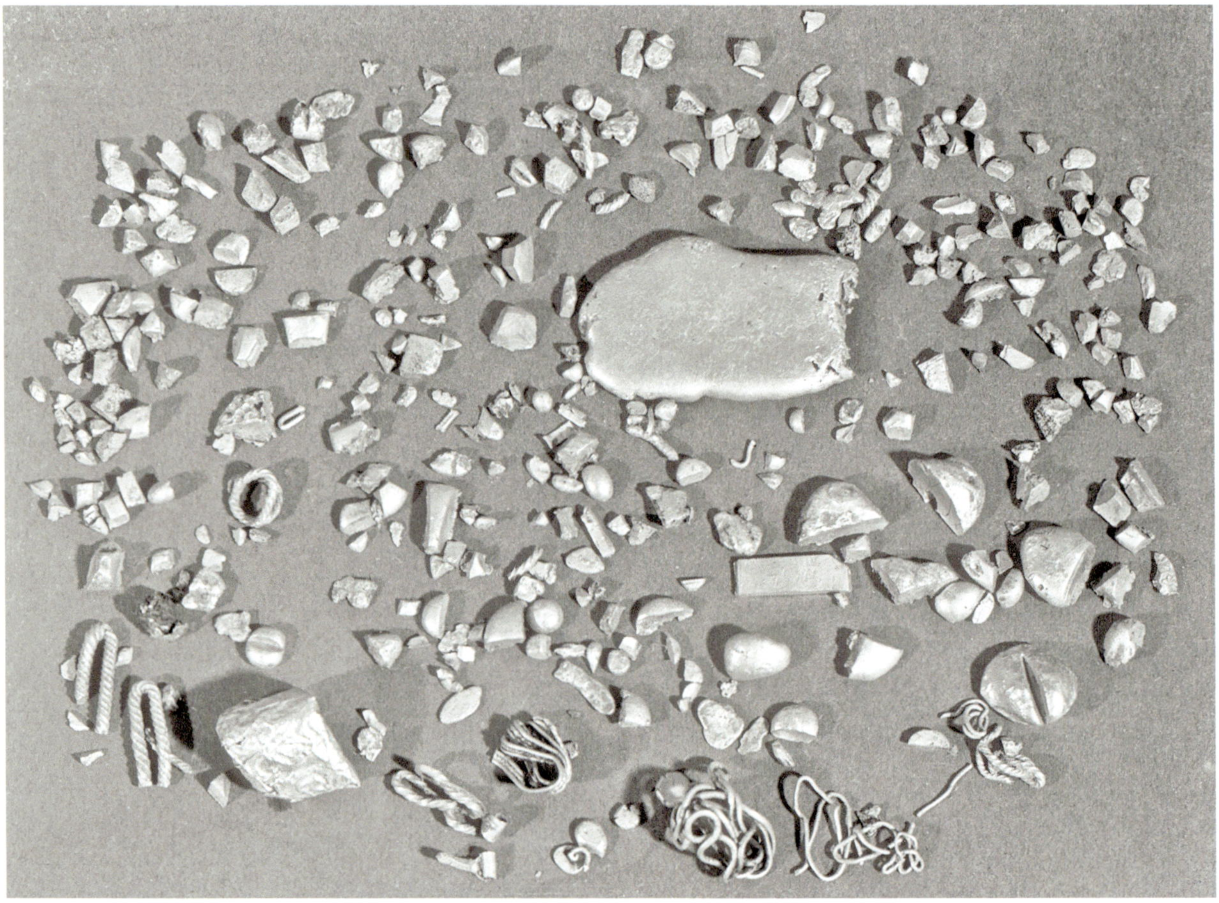

5

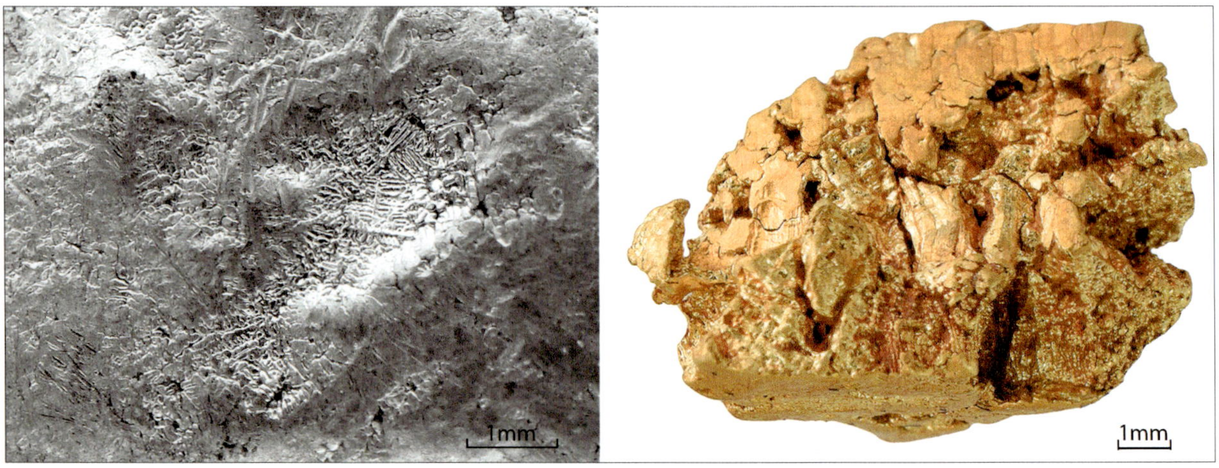

Importantly, many examples of such ingots, as well as gold nuggets, are present in the only hoard of its type – the contemporary gold hoard from Eretria (Fig. 4).[41] These ingots are often globule shaped with a slightly flat rough underside. Most of them were produced on melting plates and in many cases, dendrites are visible to the naked eye on the upper surface and in some of the cuts near the bottom surface (Fig. 5). This indicates that the ingots were not cold worked or annealed after production, but only left to cool slowly under the still burning fuel, causing the gold to solidify sufficiently slow and uneven for relatively large dendrites to form. Moreover, among several reasons, the resulting interdendritic cavities that can be observed in many cases near the bottom of the ingots made this gold unsuitable to be processed into finished objects, such as foil or jewellery, without being remelted first. This forms a marked difference with the production process suggested by the crucibles from Thessaloniki Toumba,[42] which must have been used for casting small portions of gold for further production into finished objects. In contrast, the melting plates from Methone, coupled with the evidence from Eretria, reflect an industry aimed at processing placer gold from the immediate Macedonian hinterland for the purpose of trade and supply of a maritime network.

As has been argued elsewhere, the gold in the Eretria hoard, dating to the end of the 8th century, was likely used as a form of money in transactions, being cut up where needed into smaller pieces to reach the required weight.[43] Consisting of 398 objects with a total weight of 513 gram, the objects have an average weight of 1.3 gram, but the weight distribution reflects an emphasis on very small objects, with 74% of them weighing below one gram. In fact, objects weighing between 0.1 and 0.2 gram were most numerous, suggesting these were most frequently used in transaction. As we have shown, this metrological pattern is remarkably similar to that of silver hoards from the southern Levant, dating between the late 11th and the late 8th century (Fig. 6), and indicates that the Eretria hoard is reflective of a broader phenomenon in the use and circulation of precious metal.[44] The monetary

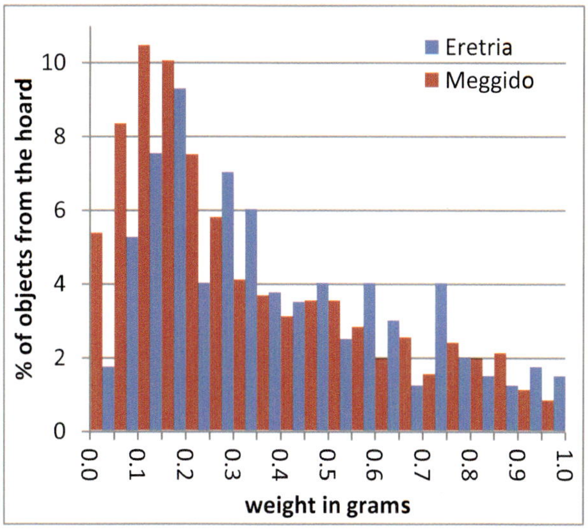

Fig. 6. Histogram of the Eretria hoard and a 10th century BCE hoard from Megiddo.

character of the hoard is furthermore confirmed by the substantial traces of wear on many of the objects, also noted in contemporary hoards from the Levant, indicating that these objects must have circulated intensively, as would be logical for money.[45]

Euboians, especially those engaged in long-distance trade, were no doubt already familiar with weighing out precious metal, as suggested among other things by the group of eastern balance weights from the 9th-century tomb 79 at the Toumba cemetery in Lefkandi.[46] The Eretria hoard though provides evidence for the use and circulation of precious metal money within a local or Aegean system. While the specific source(s) of the Eretria gold cannot at present be determined with certainty,[47] the presence of PGE (platinum Group elements) inclusions in several of the objects in the Eretria hoard (Fig. 7), together with 44 small nuggets (Fig. 8), indicate a placer source for at least part of the gold.[48] Methone, as did Sindos on the north side of the Thermaic gulf, no doubt served as a gateway for the trade in this gold, providing the Euboians – and possibly other visitors well – access to the Macedonian placer sources in its hinterland.[49]

41 Themelis 1981; 1983; Le Rider and Verdan 2002; Heymans 2021, spec. 277, with further references. We, moreover, extend our warm gratitude to Petros Themelis, for giving us permission to re-examine (t)his extraordinary discovery.

42 See note 18.

43 Verdan and Heymans 2020, 287-289; Heymans 2021, 98-101, 205-207, 277, with further references.

44 Heymans 2021, 90-101.

45 Heymans 2021, 99-100; Verdan and Heymans 2020, 289, Fig. 10.

46 Kroll 2008; Verdan and Heymans 2020, 289-290.

47 On the difficulty of determining the origin of gold by means of archaeometric analysis, see Pernicka 2014, 159-163.

48 On PGE inclusions in gold, see Ogden 1977; Craddock 2000.

49 Verdan and Heymans 2020.

7
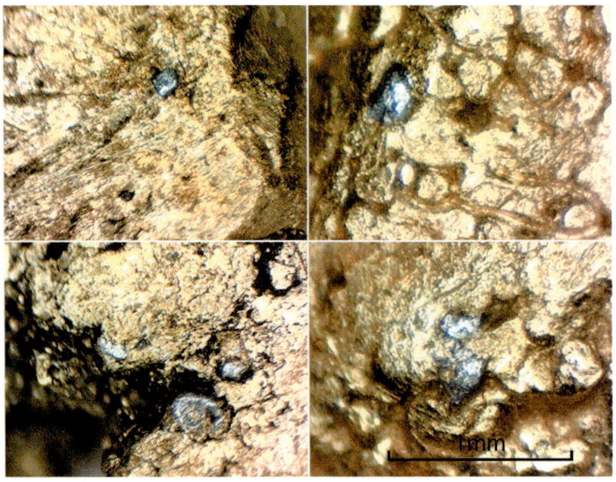

8
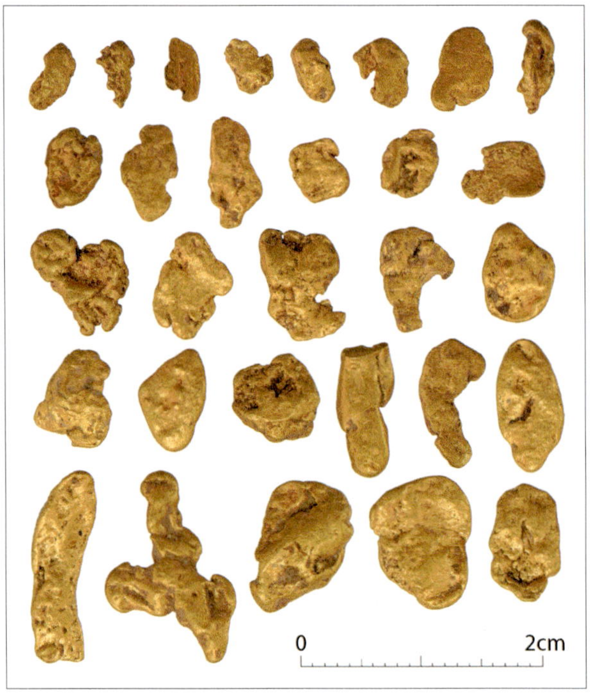

9
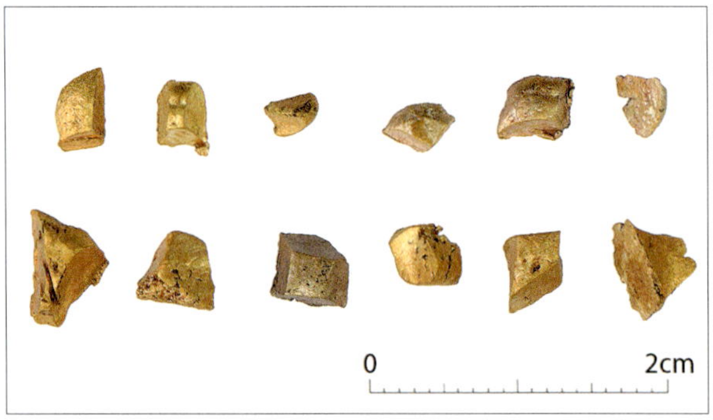

Fig. 7. Eretria hoard: PGE inclusions in gold ingots (left: Mus. Eret. inv. 14954/no. 47; right: 14952/nos 5-6; photo M. F. Guerra).

Fig. 8. Eretria hoard: selection of nuggets (photo P. Simon).

Fig. 9. Eretria hoard: colour variability of several gold ingots (photo P. Simon).

At the same time, the heterogeneity of the Eretria hoard in terms of composition (and colour: Fig. 9) of the objects (see below) suggests that we are likely dealing with the circulation of gold from different sources, possibly processed in different ways, as well as finished objects away from their primary source, as evidenced by the folded gold foil *Zungendiadem* in the hoard.[50] The exploitation of gold from the area around the Thermaic gulf should thus be considered alongside that of other sources across the northern Aegean including Mysia and even Lydia, and possibly beyond.

The changing significance of gold in Macedonia

With gold becoming high in demand as a trade commodity and Aegean currency, how were local attitudes towards gold in Macedonia affected by this increased demand? Around the same time as Euboian involvement in the exploitation and trade of northern gold was taking shape, local consumption patterns of gold appear to have shifted. Bronze was clearly still the most important metal of value, used for prestige goods such as jewellery and weapons, but simple gold objects were starting to appear as luxury items.

This can be observed in the funerary evidence from at least the 8th century, perhaps going back even to the 10th century.[51] From the Iron Age tumuli cemetery at Vergina, dating from the 10th to the 7th century, at least 391 burials have been uncovered.[52] These contained over two thousand grave goods, of which some 60% constituted metal objects, predominantly bronze jewellery and dress ornaments,[53] but including some 36 golden hair coils and 10 golden rings/ earrings found in 26 tombs.[54] Similar golden hair coils are known from at least nine other cemeteries in Macedonia, albeit in smaller numbers.[55] An

50 Diadem: see Themelis 1981, 197-200, Fig. 18-19; Themelis 1983, 163, Fig. 14.

51 See in general Chemsseddoha 2019, 248, Fig. 15.

52 Andronikos 1969; Bräuning and Kilian-Dirlmeier 2013; Chemsseddoha 2019, 380-395; 2020, 260.

53 Chemsseddoha 2019, 392, Fig. 86; 2020, 260.

54 Andronikos 1969, 259, pl. KG1–5; Bräuning and Kilian-Dirlmeier 2013, 54-56, 99, Tab. 29, 48, and Beil. 9-10; Chemsseddoha 2019, 394, Fig. 90. The exact number of objects is uncertain, because there is no table summarising the whole dataset and because the distinction between hair coils and rings has not always been clearly made.

55 In her recent survey of Iron Age funerary evidence Chemsseddoha (2019, 248, Fig. 15) reports eight other sites with golden hair coils: Kastoria-Daïlaki (*ibid.*, 299-300),

interesting example is Agrosykia, some ten kilometres north of Pella. Within a small burial group there, consisting of eleven tombs and dating mostly around the 8th century, four hair coils and a small electrum spectacle-spiral ornament were found in three tombs (one was found with no clear context). Also here, bronze grave goods were the most important class of objects, including an exquisite belt lined with bronze pendants from grave ΣΤ.[56] Presumably also from the 8th century are three graves with gold objects from an Iron Age cemetery at Karabournaki, two of which contained a pair of golden earrings, while one contained a hair coil and some pieces of foil together with Euboian-style skyphoi.[57]

While burial customs in Iron Age Macedonia were characterised by a degree of diversity,[58] gold remained rare, as illustrated by the case of Nea Philadelphia. At this extensive cemetery, in use mainly between the 9th and late 7th/early 6th century, a sheer 2.228 graves were investigated over an area of twelve acres, of which only some 40% contained grave offerings. Most of the jewellery was made of bronze or iron and only a handful of golden objects were retrieved, including a pair of earrings, and several mouthpieces.[59]

These lozenge-shaped mouthpieces made out of gold sheet point us to a new phenomenon on the horizon. They appear from the late 8th and 7th century, not only at Nea Philadelphia, but at several other sites around the Thermaic gulf,[60] and presage a surge in the use of sheet gold in graves in the region during the 6th century, most notably in the cemeteries of Archontiko and Sindos, when gold objects in several instances outnumber bronzes.[61] Golden mouthpieces from the 8th and 7th century are known also from Cyprus, Rhodes and Attica,[62] and from the Late Geometric period onwards gold foil ornaments, such as diadems, and gold grave goods in general became more widespread across the Aegean, in particular in the southern Aegean.[63] This has prompted the suggestions that gold foil grave goods in the northern Aegean were introduced at this early date by Greek colonists settling the region or appeared as a result of southern, i.e. Euboian, influence, giving a new expression to the southern Greek phenomenon of the "princely" or "heroic" burial.[64]

While Euboian colonial involvement in the area of the Thermaic gulf might have had manifold consequences, the emerging use of gold in grave offerings need not only be seen as an imported custom. Rather, local circumstances offer an important context for understanding the shifting significance of gold. Following the rather egalitarian structure of local communities during the Late Bronze Age, indications for social stratification and the emergence of a political hierarchy are becoming visible in the 9th and 8th century. At Nea Philadelphia, a large rectangular building (17.80 x 10.30 m) with various spaces and a storage function was constructed on a hill, and has been interpreted by the excavator as the residence of a local ruler.[65] And at Assiros Toumba, a large apsidal building (*c.* 15 x 8 m) was erected at the summit of the mound in Phase 1 during the 8th century.[66] It is hardly a coincidence that such developments appeared alongside an increased investment in funerary practices including the dedication of bronze and gold objects, highlighting a stronger emphasis on status and inequality.[67]

The changing use of gold thus appears to have been related to an increase in social complexity and hierarchy. However, while increasing social complexity need not have been prompted by the initial appearance of Euboian colonists on the scene,[68] both the emergence of an elite class and the role of gold in shaping and expressing their status should not be seen in isolation from the impact that gold trade likely had on the region. The visibility of structures reflecting political hierarchy in a place like

Agrosykia (*ibid.*, 348), Archontiko (*ibid.*, 352), Tzamala (*ibid.*, 376), Assomata-Tsardaki (*ibid.*, 379), Makrygialos (*ibid.*, 400), Moschochori (*ibid.*, 401), Petra-Treis Elies (*ibid.*, 407). To this list we can add Karabournaki (see text and n.57).

56 Chrysostomou 1991; Chrysostomou *et al.* 2007.

57 Panti and Vasileiou 2018, 105, 107, 110-111, 120, 131.

58 Chemsseddoha 2017, 389-390; 2019.

59 Misailidou-Despotidou 2008, 36-40.

60 Despini 2009, 34-35; Saripanidi 2017, 106; Chemsseddoha 2019, 248 (Fig. 15) and 259-262 (Fig. 19).

61 Chrysostomou and Chrysostomou 2012; Saripanidi 2017, 81-89, 134, Fig. 7; 2019; Despini 2016; Hatzopoulos 2020, 19-23. For the cemetery near Thessaloniki Toumba, see Chavela 2022, 56-58.

62 Saripanidi 2017, 98-99, 106-107.

63 Gold foil ornaments: Ohly 1953. Eretria: Blandin 2007, vol. 1, 92-98. Increased popularity of gold grave goods in the southern Aegean was no doubt also stimulated by Phoenician contacts with Crete and the Dodecanese, although the question where this gold came from remains unexplored.

64 Despini 2009, 35-38; Saripanidi 2017, 106-108.

65 Misailidou-Despotidou 2008, 28-34.

66 Wardle and Wardle 2000.

67 Andreou 2020, 928-930

68 Gimatzidis 2020.

Nea Philadelphia, an upstream settlement along the gold-bearing Gallikos river, certainly counts as an indication for this: this settlement, no doubt active in the exploitation of placer gold, flourished between the 9th and 7th century as a result of the gold trade.

Euboian demise?

After the early 7th century, evidence for Euboian colonial and maritime activity decreases. This has often been explained with reference to the socalled Lelantine War, which would have exhausted its main players, Chalcis and Eretria, although the historicity of this conflict has been contested.[69] After the start of the century, the export of Euboian pottery, the main indicator of Euboian maritime activity, appears to have dropped but decreased visibility of Euboian pottery in the archaeological record need not indicate an abrupt downturn. Firstly, identifying Euboian vases from this period is complicated by the fact that they have fewer distinctive features than in the Late Geometric period. Secondly, local production shows a decline in quality, in terms of both fabric and decoration, explaining why it became less interesting for export. This is particularly the case for banquet vessels (craters, drinking vases), which had previously circulated in large numbers. Yet, if we accept the evidence as representative of a downturn, it would also reflect a dearth in long-distance maritime activity: Euboian pottery disappeared in the Levant mostly after 700, and even at Al Mina, East Greek pottery appears to have replaced Euboian pottery by 670 (level 6) at the latest.[70] Moreover, other indicators also signal the loss of Euboian colonial supremacy. No new Euboian colonies were established in the region after the early 7th century (with the possible exception of Dikaia)[71] and Euboians were superseded by other colonisers, such as Andrians, Parians and Corinthians.[72] These changes were no doubt the result of multiple and complex factors, and we can hardly aim to consider these here in full. We do, however, want to address one potential aspect of this perceived demise.

While the Gallikos river was still referred to in the 5th century by Herodotus as the *Echedoros* – gift bearing, no ancient source makes an actual mention of its gold content.[73] It could well be the case that large-scale exploitation of placer sources around the Thermaic gulf had become less profitable in the course of the 7th century and that the Gallikos was no longer actively exploited in later times, with only the name reflecting its former wealth. Euboian demand for gold over the course of two centuries or more had stimulated exploitation by local groups, also testified by the prospering of local settlements upstream such as Nea Philadelphia. Especially since large-scale exploitation before the Iron Age is uncertain if not unlikely, efforts to exploit the gold must have been very profitable at first. However, natural resources are not historically static and replenishment through erosion or geological events could most likely not keep up with exploitation. As a result, these placer sources were bound to become exhausted beyond profitability at some point.[74] Whether that point was reached for the Euboians during the 7th century is hard to tell. While new evidence for melting plates at Methone from the Archaic workshop area (c. 675-550) suggests that gold processing could have continued at least into the 7th century,[75] it is possible that the yield of the deposits diminished and that Euboian traders lost interest as silver replaced gold as the dominant Aegean commodity of value and currency in the later 7th century.[76] It is also possible that exploitation continued but that they simply lost any preferential access to the gold they might have had. The prominence of gold in wealthy burials of the 6th century in Archontiko and Sindos could indicate continuity in local exploitation, although this gold could have come from other primary and placer gold deposits across the northern Aegean as well.[77] In any event, it is more relevant to note in this regard that in order to produce large pieces of foil only small bits

69 Parker 1997, 91-92; Coldstream 2003, 182; Walker 2004, 89-92, 161-163; cf. Hall 2007, 1-8; Verdan and Heymans 2020, 291.

70 Fantalkin 2006, 201; Vacek 2017.

71 Voutiras 2008 (dating the foundation after 470); *contra* Tiverios 2013, 100 (8th century).

72 There is evidence that (local) Chalkidians joined others, such as the Andrians in their foundations of Sane and Akanthos, but this is usually considered as a sign of weakness from the Chalkidians. See Plut. *Quaest. Graec.* 30; Graham 1978, 96; Tiverios 2008, 52-66.

73 Hdt. 7.124, 127; cf. Skyl. 66; Apollod. 2.5.11.

74 Several sources in fact note the exhausted gold mines. When touching upon the gold mines of the ruined city of Astyra, in the Troad, Strabo (13.1.23) notes that "these mines are now scant, being used up, like those on Mt. Tmolus in the neighbourhood of the Pactolus River." Elsewhere (14.5.28), he relates that the Lydian kings got their wealth from mines "in Lydia and from the region between Atarneus and Pergamum, where is a small deserted town, whose lands have been exhausted of ore." (transl. Loeb).

75 We thank Sarah Morris and John Papadopoulos for generously sharing this information with us. On the Archaic workshop area, see Morris *et al.* 2020, 699-702.

76 Cf. Heymans 2021, 217-225.

77 Cf. Vavelidis and Andreou 2008, 362, Fig. 1, cf. Saripanidi 2019, 181.

of gold are needed, and that the gold in these burials thus cannot possibly compare to the quantities that had been traded earlier by the Euboians, such as that preserved in the Eretria hoard.

A wider phenomenon

Now, should the Euboian endeavour to secure access to gold resources be regarded as an isolated phenomenon? Euboians might have been the first but they were certainly not the only group to take an active interest in northern gold sources in the early Greek world. As suggested earlier, Euboian involvement in regions around the Thermaic gulf can be understood in a broader context of emerging efforts to exploit gold resources and the increase of its use within the Aegean.

A starting point here is to look again at the Eretria hoard. As noted, the composition of the objects in the hoard is diverse, with gold ratios varying between 99.6% and 45.5%.[78] Considering the evidence for the intensive circulation of the objects, could it be that the hoard contained gold from diverse provenances, not unlike a later mixed coin hoard? The hoard could have served to carry out transactions of the sort that took place in Methone, involving traders from across the Aegean and beyond, and thereby potentially absorbed metal from various sources.

Unminted gold objects, typologically similar to the globule and token (flattened globules) ingots in the Eretria hoard, and dating at least from the third quarter of the 7th century are known also from the Artemision at Ephesus, which was an Aegean seaport. Significantly, these were found together with the first minted coins.[79] Slightly later, dating to the late 7th or early 6th century, an inscribed lead tablet from the sanctuary preserves a record of the sanctuary's financial administration, indicating the use of unminted gold and silver in maritime trade.[80]

Was Euboian exploitation of gold and adoption of gold money in the context of trade paralleled by other players in the Aegean? Little direct evidence survives of an early Phoenician involvement or presence at Thasos. Herodotus relates that the Phoenicians were the first to exploit primary gold deposits on the island,[81] and John Graham suggested more than fifty years ago that Phoenicians exploited the metal sources on the island before a crisis in Phoenicia proper following the destruction of Sidon in 677 by the Assyrians led them to retreat, after which Parian colonists came to settle the island.[82] Much later, Pliny the Elder records that gold mining and smelting would have been invented by the Phoenician Kadmos at Pangaion, and also Strabo associates Kadmos with the precious metal sources in the region, noting that he gained his wealth from the mines "around Thrace and mount Pangaion."[83] One wonders why these traditions focus specifically on gold, if the exploitation of silver sources on Thasos and the area of the Pangaion was certainly no less important and, moreover, had considerable antiquity, with evidence for cupellation at Limenaria on Thasos dating already to the late fourth millennium.[84]

An answer to this question could be sought in the fact that until the mid-7th century silver was presumably not yet exploited on a significant scale. Industrial production of silver demanded a much higher investment and bigger scale of organisation in order to become profitable, while exploitation of primary and placer gold could be profitable on a relatively small scale. The amount of labour and fuel needed for mining and the processing of argentiferous lead ores in order to produce silver on an industrial scale presumes a serious level of investment and organisation. On the contrary, gold exploitation requires no complex logistics and can be undertaken on a small scale with a proportionate yield.[85] Could it be that these traditions

78 XRF analyses were carried out in the framework of our ongoing study of the hoard. This project is undertaken in cooperation with Maria Filomena Guerra, whom we thank warmly for her invaluable help and input, and would not have been possible without the support of Angeliki Simosi, head of the Ephorate of Antiquities of Euboia, and Athena Athanassiadou, of the Ephorate of Antiquities of Pieria, whom we gratefully acknowledge. We are furthermore thankful for the support provided by the Swiss School of Archaeology in Greece and the Mediterranean Archaeological Trust.

79 Hogarth 1908, 119; Robinson 1951, 156, 166-167; Kerschner and Konuk 2020, 94-95, 102, 107-108, 111-113, figs 5.1, 7.83, 8.A-G, 9.

80 Kroll 2020.

81 Hdt. 6.47.

82 Graham 1978; cf. Morris 1992, 144-145; Tiverios 2008, 74-76, 126; Sanidas *et al.* 2018.

83 Plin. *HN* 7.197; Strab. 14.5.28.

84 E.g. Bassiakos *et al.* 2019.

85 Borg 2014. Non-industrial mining, generally referred to artisanal and small-scale mining (ASM) is common still today and most widespread in developing countries. Artisanal and small-scale gold mining accounts for 20% of the global supply. See https://www.worldbank.org/en/topic/extractiveindustries/brief/artisanal-and-small-scale-mining#:~:text=Artisanal%20and%20Small%2DScale%20Mining%20occurs%20in%20approximately%2080%20countries,to%2020%25%20of%20diamond%20mining; ASGM report 2022 – World Gold Council (https://www.gold.org/esg/artisanal-and-small-scale-gold-mining#registration-type=google&just-verified=1).

emphasising gold reflect memories of a period – in the 8th and early 7th century – during which silver did not yet function as an important currency and trade commodity and Phoenicians maintained access to the region's gold?

Recent research now also highlights Lydian interest in gold sources in northwestern Asia Minor – Mysia, Aeolis and the Troad – from at least the early 7th century on. Under the early Mermnad dynasty, Lydia expanded its presence into these regions, with the likely purpose of controlling its gold sources.[86] Its ruler, Gyges, was referred to by his contemporary, Archilochos, as "rich in gold" and Strabo would later note that Gyges (and Alyattes and Kroisos) based his wealth on the mines in Lydia and in the region between Atarneus and Pergamon.[87]

Gold must have been of good use to the Lydians already in the early 7th century and, shortly after, this led them to produce their famous coinage. While it was long thought that Lydian royal issues were produced from natural electrum in the Pactolus river, it is now clear that these coins were artificially manipulated and the supply of gold from the much wealthier sources of the northwestern Aegean must have played an important role in their production.[88] So, arguably, what appears to have started with Euboian interest in gold sources around the Thermaic gulf can be understood in a broader context of exploitation and use of gold in the northern Aegean, including – in the second half of the 7th century – the introduction of coinage in Lydia.

Conclusion

The purpose of this article was to show how the significance and value of gold changed between the 12th and the 10th century, and again between the 10th and the 7th century for the different players involved in the exploitation and trade of gold in the northern Aegean. At the same time, tracing the role of gold allowed us to see how its sources and the interactions revolving around it shaped the different societies across the colonial environment, while in turn, broader developments in these societies gave rise to changing regimes of value and the appropriation of gold in that context. While this article is not primarily concerned with colonial interactions, it is clear that the interplay between local communities, colonial groups and the environment was key to the changing role of gold. Taking explicit account of these three factors, we can outline our understanding of this development in the following way.

The presence of placer gold sources in the region around the Thermaic gulf formed one of the reasons that Euboians were drawn to the area early on. Local societies of the Late Bronze Age in central Macedonia appear to have had conspicuously little use for the precious metal as a status marker, because their social structures afforded limited space for emphasising inequality. In Euboia, however, gold played an important role as a prestige good and contacts between early Euboian visitors/settlers and local groups in the north provided a way to acquire gold, as communities started to systematically exploit placer deposits. Increased productivity, paralleled by the intensification of interactions revolving around the exploitation of gold, including the establishment of a Euboian colonial presence, in turn enabled the adoption of gold as a form of money by the 8th century.

Having secured access to gold sources through their colonial presence in the region and local contacts, Euboians enjoyed the benefits of its use in Aegean trade. At the same time, increasing social complexity in local communities impacted by the colonial environment went hand in hand with the appropriation of gold within local Macedonian regimes of value, a development that would continue into the Archaic period. In this way, long-term interactions between Euboians, local groups and their environment ended up shaping both local attitudes towards gold, and the changing role of the metal among Euboians.

This was not an isolated phenomenon, and the adoption of gold as an Aegean form of money was likely contingent upon the exploitation and use of gold by other players in the Aegean. Although limited evidence is available, other gold sources in the northern Aegean – in Thrace, Mysia and the Troad, appear to have become the focus of interest by the 8th and early 7th century, as electrum became the preferred metal of value in the Aegean.

86 Cahill *et al.* 2020, 312–320.

87 Archil. fr. 19; Strab. 14.5.28. See Cahill *et al.* 2020, 317, for additional references.

88 Cahill *et al.* 2020, 312–320; Van Alfen and Wartenberg 2020, 4-5.

Bibliography

S. Andreou, "Exploring the patterns of power in the Bronze Age settlements of northern Greece", in: K. Branigan (ed.), *Urbanism in the Aegean Bronze Age*, London, 2001, 160-173.

S. Andreou, "Northern Aegean", in: E. H. Cline (ed.), *The Oxford Handbook of the Bronze Age Aegean (ca. 3000-1000 BC)*, New York and Oxford, 2010, 643-659.

S. Andreou, "The Thermaic gulf", in: I. Lemos and A. Kotsonas (eds), *A Companion to the Archaeology of Early Greece and the Mediterranean*, Hoboken NJ, 2020, 913-938.

S. Andreou and K. Efkleidou, "Η πανεπιστημιακή ανασκαφή στη Τούμπα Θεσσαλονίκης, 2008", *Το Αρχαιολογικό Έργο στη Μακεδονία και τη Θράκη* 22, (2011), 323-328.

S. Andreou and M. Vavelidis, "So rich and yet so poor: Investigating the scarcity of gold artefacts in Bronze Age northern Greece", in: H. Meller, R. Risch and E. Pernicka (eds), *Metalle der Macht – Frühes Gold und Silber. Metals of Power – Early Gold and Silver, 6. Mitteldeutscher Archäologentag, 17-19. Oktober 2013*, Halle, 2014, 451-465 [*Tagungen des Landesmuseum für Vorgeschichte Halle* 12].

M. Andronikos, *Βεργίνα Ι. Το Νεκροταφείον των Τύμβων*, Athens, 1969 [*Βιβλιοθήκη της εν Αθήναις Αρχαιολογικής Εταιρείας* 62].

Y. Bassiakos, N. Nerantzis and S. Papadopoulos, "Late Neothic/Early Bronze Age metallurgical practices at Limenaria, Thasos: Evidence for silver and copper production", *Archaeological and Anthropological Sciences* 11/6 (2019), 2743-2757.

M. Bessios, I. Tzifopoulos and A. Kotsonas, *Μεθώνη Πιερίας Ι. Επιγραφές, χαράγματα και εμπορικά σύμβολα στη γεωμετρική και αρχαϊκή κεραμική από το "Υπόγειο" της Μεθώνης Πιερίας στη Μακεδονία*, Thessaloniki, 2012.

B. Blandin, *Les pratiques funéraires d'époque géométrique à Erétrie. Espace des vivants, demeures des morts*, Gollion, 2007 [*Eretria* XVII].

G. Borg, "Gold is where you find it – Zeitgenössischer artisanaler Goldbergbau in Afrika als Analogie (prä-) historischer Goldgewinnung", in: H. Meller, R. Risch and E. Pernicka (eds), *Metalle der Macht – Frühes Gold und Silber. Metals of Power – Early Gold and Silver, 6. Mitteldeutscher Archäologentag, 17-19. Oktober 2013*, Halle, 2014, 53-69 [*Tagungen des Landesmuseum für Vorgeschichte Halle* 12].

A. Bräuning and I. Kilian-Dirlmeier, *Die eisenzeitlichen Grabhügel von Vergina. Die Ausgrabungen von Photis Petsas 1960-1961*, Mainz, 2013 [*Monographien des Römisch-Germanischen Zentralmuseums* 119].

N. Cahill, J. Hari, B. Önay and E. Dokumaci, "Depletion gilding of Lydian electrum coins and the sources of Lydian gold", in: P. Van Alfen, U. Wartenberg, W. Fischer-Bossert, H. Gitler, K. Konuk and C. C. Lorber (eds), *White Gold: Studies in early electrum coinage*, New York and Jerusalem, 2020, 291-336.

X. Charalambidou, "Viewing Euboea in relation to its colonies and relevant sites in northern Greece and Italy", in: A. Mazarakis Ainian, A. Alexandridou and X. Charalambidou (eds), *Regional Stories Towards a New Perception of the Early Greek World, Acts of an International Symposium in honour of Professor Jan Bouzek, Volos 18-21 June 2015*, Volos, 2017, 85-119.

K. Chavela, "Transformations and formations around the Thermaic gulf in the Late Bronze and Early Iron Age: The evidence of burial practices", in: S. Gimatzidis, M. Pieniazek and S. Mangalaglou-Vortuba (eds), *Archaeology across Frontiers and Borderlands: Fragmentation and Connectivity in the North Aegean and the Central Balkans from the Bronze Age to the Iron Age*, Vienna, 2018, 159-186.

K. Chavela, "'Poor' indigenous and 'wealthy' Macedonians (?). The evidence of burial practices around the Thermaic gulf (Thessaloniki)", in: U. Kelp and W.-R. Teegen (eds), *Wealthy and Healthy? Methodological Approaches to Non-Élite Burials, Proceedings of the 19th International Congress of Classical Archaeology, Cologne/Bonn 22 – 26 May 2018, Archaeology and Economy in the Ancient World, Panel 1.2*, Heidelberg, 2022, 53-70.

A.-Z. Chemsseddoha, "New perspectives on the burial customs in Macedonia during the Early Iron Age", in: A. Mazarakis Ainian, A. Alexandridou and X. Charalambidou (eds), *Regional Stories Towards a New Perception of the Early Greek World, Acts of an International Symposium in honour of Professor Jan Bouzek, Volos 18-21 June 2015*, Volos, 2017, 381-394.

A.-Z. Chemsseddoha, *Les pratiques funéraires de l'âge du Fer en Grèce du Nord: étude d'histoires régionales*, Bordeaux, 2019 [*Scripta Antiqua* 121].

A.-Z. Chemsseddoha, "Connections between the Balkans and the Aegean", in: M. Gavranović, D. Heilmann, A. Kapuran and M. Verčík (eds), *Spheres of Interaction: Contacts and Relationships between the Balkans and Adjacent Regions in the Late Bronze / Iron Age (13th–5th Centuries BCE)*, Rahden/ Westf., 2020, 253-269.

A. Chrysostomou, "Νεκροταφείο της εποχής Σιδήρου στην Αγροσυκιά Γιαννιτσών", *Το Αρχαιολογικό Έργο στη Μακεδονία και τη Θράκη* 5 (1991), 127-136.

A. Chrysostomou and P. Chrysostomou, "The "gold-wearing" Archaic Macedonians from the western cemetery of Archontiko, Pella", in: M. Tiverios, P. Nigdelis and P. Adam-Veleni (eds), *Θρεπτήρια: μελέτες γία την αρχαία Μακεδονία, Threpteria: Studies on Ancient Macedonia*, Thessaloniki, 2012, 491-516.

P. Chrysostomou, I. Aslanis and A. Chrysostomou, *Αγροσυκιά: ένας οικισμός των προϊστορικών και ιστορικών χρόνων*, Veria, 2007.

J.N. Coldstream, *Geometric Greece, 900-700 B.C.*, London, 2003 [1977].

J.N. Coldstream, "The Rich Lady of the Areiopagos and her contemporaries: A tribute in memory of Evelyn Lord Smithson", *Hesperia* 64/4 (1995), 391-403.

P.T. Craddock, "The Platinum Group Element Inclusions", in: A. Ramage and P.T. Craddock and Craddock (eds) *King Croesus' Gold. Excavations at Sardis and the History of Gold Refining*, Cambridge, 2000, 238-244 [*Archaeological Exploration of Sardis* 11].

J.P. Crielaard, "Basileis at sea: Elites and external contacts in the Euboean gulf region from the end of the Bronze Age to the beginning of the Iron Age", in: S. Deger-Jalkotzy and I.S. Lemos (eds), *Ancient Greece from the Mycenaean Palaces to the Age of Homer*, Edinburgh, 2006, 271-297 [*Edinburgh Leventis Studies* 3].

J.P. Crielaard, "*Hygra keleutha*. Maritime matters and the ideology of seafaring in the Greek epic tradition", in: *Alle origini della Magna Grecia. Mobilità, migrazioni, fondazioni, Atti del cinquantesimo convegno di studi sulla Magna Grecia, Taranto 1-4 ottobre 2010*, Taranto, 2012, 135-157.

J.-P. Descoeudres, "Central Greece on the eve of the colonisation movement", in: G.R. Tsetskhladze (ed.), *Greek Colonisation: An Account of Greek Colonies and Other Settlements Overseas*, vol. 2, Leiden and Boston, 2008, 289-382.

A. Despini, "Gold funerary masks", *AntK 52* (2009), 20-65.

A. Despini, *Σίνδος ΙΙΙ: το νεκροταφείο, ανασκαφικές έρευνες 1980-1982. Μάσκες και χρυσά ελάσματα, κοσμήματα, μικροαντικείμενα και στλεγγίδες, είδη οπλισμού*, Athens, 2016 *[Βιβλιοθήκη της εν Αθήναις Αρχαιολογικής Εταίρειας 309]*.

N. Erb-Satullo, "Technological rejection in regions of early gold innovation revealed by geospatial analysis", *Scientific Reports* 11 (2021), https://doi.org/10.1038/s41598-021-98514-7.

A. Fantalkin, "Identity in the making: Greeks in the eastern Mediterranean during the Iron Age", in: A. Villing and U. Schlotzhauer (eds), *Naukratis: Greek Diversity in Egypt. Studies on East Greek Pottery and Exchange in the Eastern Mediterranean*, London, 2006, 199-208.

R. Firth, *Primitive Polynesian Economy*, London, 1965 [1939].

T.F. Garrard, *Akan Weights and the Gold Trade*, London, 1980.

S. Gimatzidis, *Die Stadt Sindos: eine Siedlung von der Späten Bronze- bis zur Klassischen Zeit am Thermaischen Golf in Makedonie*n, Rahden/Westf, 2010 [*Prähistorische Archäologische in Südosteuropa* 26].

S. Gimatzidis, "Counting sherds at Sindos: Pottery consumption and construction of identities in the Iron Age", in: S. Verdan, T. Theurillat and A. Kenzelmann Pfyffer (eds), *Early Iron Age Pottery: A Quantitative Approach, Proceedings of the International Round Table, Athens 28-30 November 2008*, Oxford, 2011, 97-110 [*BAR IS* 2254].

S. Gimatzidis, "The economy of early Greek colonisation in the northern Aegean", *Journal of Greek Archaeology* 5 (2020), 243-262.

S. Gimatzidis, "Early Greek Colonisation in the Northern Aegean: A New Perspective from Mende", in: C. Colombi, V. Parisi, O. Dally, M. A. Guggisberg and G. Piras (eds), *Comparing Greek Colonies: Mobility and Settlement Consolidation from Southern Italy to the Black Sea (8th-6th Century BC), Proceedings of the International Conference (Rome, 7-9.11.2018)*, Berlin, 2022, 52-67

A.J. Graham, "The foundation of Thasos", *BSA* 73 (1978), 61-98.

J.M. Hall, *A History of the Archaic Greek World, ca. 1200–479 BCE*, Chichester, 2007.

M.B. Hatzopoulos, *Ancient Macedonia*, Berlin and Boston, 2020.

E.W. Herbert, *Red Gold of Africa: Copper in Precolonial History and Culture*, Madison and London, 1984.

E.D. Heymans, *The Origins of Money in the Iron Age Mediterranean World*, Cambridge, 2021.

R.A. Higgins, "The jewellery", in Popham *et al.* 1979-1980, vol. text, 1980, 217-225.

A. Hochstetter, *Kastanas. Ausgrabungen in einem Siedlungshügel der Bronze- und Eisenzeit Makedoniens 1975-1979. Die Kleinfunde*, Berlin, 1987 [*Prähistorische Archäologie in Südosteuropa* 6].

D.G. Hogart, *Excavations at Ephesos. The Archaic Artemisia*, London, 1908.

P. Ilieva, "Phoenicians, Cypriots and Euboeans in the northern Aegean: A reappraisal", *Aura* 2 (2019), 65-102.

R.E. Jones 1980, "Analyses of gold objects from the cemeteries", in: Popham *et al.* 1979-1980, vol. text, 461-464.

G. Karo, *Die Schachtgräber von Mykenai*, Munich, 1930.

A. Kasseri, "Φοινικικοί εμπορικοί αμφορείς από τη Μεθώνη Πιερίας", in: E. Kefalidou and D. Tsiafaki (eds), *Κεραμέως παίδες. Αντίδωρο στον Καθηγητή Μιχάλη Τιβέριο από τους μαθητές του*, Thessaloniki, 2012, 299-308.

M. Kerschner and K. Konuk, "Electrum coins and their archaeological context: The case of the Artemision of Ephesus", in: P. Van Alfen, U. Wartenberg, W. Fischer-Bossert, H. Gitler, K. Konuk and C. C. Lorber (eds), *White Gold: Studies in Early Electrum Coinage*, New York and Jerusalem, 2020, 83-190.

A. Kotsonas, "What makes a Euboean colony or trading station? Zagora in the Cyclades, Methone in the Thermaic gulf, and Aegean networks in the 8th century BC", in: J.-P. Descoeudres and S.A. Paspalas (eds), *Zagora in Context: Settlements and Intercommunal Links in the Geometric Period (900-700 BC), Proceedings of the Conference, Athens, 20-22 May 2012*, *MeditArch* 25 (2012), 243-257.

J.H. Kroll, "Early Iron Age balance weights at Lefkandi, Euboea", *OJA 27*, 1 (2008), 37-48.

J.H. Kroll, "The inscribed account on lead from the Ephesian Artemisium", in: P. Van Alfen, U. Wartenberg, W. Fischer-Bossert, H. Gitler, K. Konuk and C. C. Lorber (eds), *White Gold: Studies in Early Electrum Coinage*, New York and Jerusalem, 2020, 49-64.

T.J. LeCain, "Environmental history and global mining: Towards a neo-materialist approach", in: S. Berger and P. Alexander (eds), *Making Sense of Mining History*, London, 2019, 283-301.

I.S. Lemos, "The Lefkandi connection: Networking in the Aegean and the eastern Mediterranean", in: L. Bonfante and V. Karageorghis (eds), *Italy and Cyprus in Antiquity: 1500-450 BC, Proceedings of an International Symposium held at the Italian Academy for Advanced Studies in America at Columbia University, November 16- 18, 2000*, Nicosia, 2001, 215-226.

I.S. Lemos, *The Protogeometric Aegean. The Archaeology of the Late Eleventh and Tenth Centuries BC*, Oxford, 2002.

I.S. Lemos, "Euboea", in: I.S. Lemos and A. Kotsonas (eds), *A Companion to the Archaeology of Early Greece and the Mediterranean*, Hoboken NJ, 2020, 787-813.

B. Leone, "A Trade Route between Euboea and the Northern Aegean", in: J.-P. Descoeudres and S. A. Paspalas (eds), *Zagora in Context: Settlements and Intercommunal Links in the Geometric Period (900–700 BC), Proceedings of the Conference, Athens, 20–22 May 2012*, *MeditArch* 25 (2012), 229-241.

G. Le Rider and S. Verdan, "La trouvaille d'Erétrie: réserve d'un orfèvre ou dépôt monétaire?", *AntK* 45 (2002), 133-152.

C. Martin Pruvot, K. Reber and T. Theurillat, *Cité sous terre: des archéologues suisses explorent la cité grecque d'Érétrie*, Gollion, 2010.

K.R. Maxwell-Hyslop, *Western Asiatic Jewellery c.3000-612 B.C.*, London, 1971.

N.D. Meeks and P.T. Craddock, "Scientific examination of two sherds with gold particles", in Verdan 2013, 271-273.

V. Misailidou-Despotidou, "Νέα Φιλαδέλφεια, οικιστικές εγκαταστάσεις και νεκροταφεία στην ενδοχώρα της Θεσσαλονίκης", in: G. Aikaterinidis (ed.), *Αρχαιολογικές τροχιοδρομήσεις από τη Θεσσαλονίκη στον Πλαταμώνα*, Athens, 2008, 24-69.

S.P. Morris, *Daidalos and the Origins of Greek Art*, Princeton, 1992.

S.P. Morris and J.K. Papadopoulos (eds), *Ancient Methone 2003-2011*, 2 vol., Los Angeles, 2023 [*Monumenta Archaeologica* 49].

S.P. Morris, J.K. Papadopoulos, M. Bessios, A. Athanassiadou and K. Noulas, "The Ancient Methone Archaeological Project: A preliminary report on fieldwork, 2014-2017", *Hesperia* 89 (2020), 659-723.

S. Moschonissioti, "Excavations at ancient Mende", in: M. Bats and B. D'Agostino (eds), *Euboica: L'Eubea e la presenza euboica in Calcidica e in Occidente, Atti del Convegno Internazionale di Napoli, 13-16 Novembre 1996*, Naples, 1998, 255-271.

S. Moschonissioti, "The city of Mende during the late eighth and seventh centuries BC", in: X. Charalambidou and C. Morgan (eds), *Interpreting the Seventh Century BC: Tradition and Innovation, Proceedings of the International Colloquium Conference held at the British School at Athens, 9th-11th December 2011*, Oxford, 2017, 355-363.

J. M. Odgen, "Platinum Group metal inclusions in ancient gold artifacts", *Journal of the Historical Metallurgy Society*, 11/2 (1977), 53-72.

D. Ohly, *Griechische Goldbleche des 8. Jahrhunderts v. Chr.*, Berlin, 1953.

A. Panti and S. Vasileiou, "Το νεκροταφείο της ΠΕΣ στην οδό Θ. Σοφούλη στο Καραμπουρνάκι Θεσσαλονίκης", *ArchDelt* 69-70, Α' 2014-2015 (2018), 83-144.

J.K. Papadopoulos, "Euboians in Macedonia? A Closer Look", *OJA* 15/2 (1996), 151-181.

J.K. Papadopoulos, "Phantom Euboians – A decade on", in: D.W. Rupp and J.E. Tomlinson (eds), *Euboea and Athens, Proceedings of a Colloquium in Memory of Malcolm B. Wallace, Athens 26-27 June 2009*, Toronto, 2011, 113-133.

V. Parker, *Untersuchungen zum lelantischen Krieg und verwandten Problemen der frühgriechischen Geschichte*, Stuttgart, 1997.

E. Pernicka, "Possibilities and limitations of provenance studies of ancient silver and gold", in: H. Meller, R. Risch and E. Pernicka (eds), *Metalle der Macht – Frühes Gold und Silber. Metals of Power – Early Gold and Silver, 6. Mitteldeutscher Archäologentag, 17-19. Oktober 2013*, Halle, 2014, 153-164 [*Tagungen des Landesmuseum für Vorgeschichte Halle* 12].

M.R. Popham, "Precolonization: Early Greek contact with the East", in: G.R. Tsetskhladze and F. De Angelis (eds), *The Archaeology of Greek Colonisation: Essays Dedicated to Sir John Boardman*, Oxford, 1994, 11-34.

M.R. Popham and I.S. Lemos, *Lefkandi III. The Toumba Cemetery: The excavations of 1981, 1984, 1986 and 1992-1994, Plates*, 1996, London [*BSA* Suppl. 29].

M.R. Popham, L.H. Sackett and P.G. Themelis (eds), 1979-1980, *Lefkandi I. The Iron Age: The Settlement, The Cemeteries*, 2 vol., London [*BSA* Suppl. 11].

M.R. Popham, E. Touloupa and L.H. Sackett, "Further excavation of the Toumba cemetery at Lefkandi, 1981", *BSA* 77 (1982), 213-248.

M.R. Popham, P. Calligas and L.H. Sackett, "Further excavation of the Toumba cemetery at Lefkandi, 1984 and 1986: A preliminary report", *AR* 35 (1988-1989), 117-129.

A. Ramage and P.T. Craddock, *King Croesus' Gold. Excavations at Sardis and the History of Gold Refining*, Cambridge, 2000 [*Archaeological Exploration of Sardis* 11].

E.S.G. Robinson, "The coins from the Ephesian Artemision reconsidered", *JHS* 71 (1951), 156-167.

G.M. Sanidas, D. Malamidou and N. Nerantzis, "Paros, Thasos and the question of mining and metal production before and after colonisation: Review and perspectives", in: D. Katsonopoulou (ed.), *Paros and its Colonies, Proceedings of the Fourth International Conference on the Archaeology of Paros and the Cyclades, Paros, 11-14 June 2015*, Athens, 2018, 251-265.

V. Saripanidi, "Constructing continuities with a "heroic" past: Death, feasting and political ideology in the Archaic Macedonian kingdom", in: A. Tsingarida and I.S. Lemos (eds), *Constructing Social Identities in Early Iron Age and Archaic Greece*, Brussels, 2017, 73-170 [*Études d'archéologie* 12].

V. Saripanidi, "Macedonian necropoleis in the Archaic period: Shifting practices and emerging identities", in: H. Frielinghaus, J. Stroszeck and P. Valavanis (eds), *Griechische Nekropolen: Neue Forschungen und Funde*, Möhnesee, 2019, 175-196.

A. Simosi, *Αρέθουσα. Το Αρχαιολογικό Μουσείο Χαλκίδας*, Athens, 2022.

A.M. Snodgrass, "The Euboeans in Macedonia: A new precedent for westward expansion", in: B. D'Agostino and D. Ridgway (eds), *Apoikia. I più antichi insediamenti greci in Occidente : funzioni e modi dell'organizzazione politica e sociale, Scritti in onore di Giorgio Buchner*, Naples, 1994, 87-93.

Z.A. Stos-Gale, "Silver vessels in the Mycenaean shaft graves and their origin in the context of the metal supply in the Bronze Age Aegean", in: H. Meller, R. Risch and E. Pernicka (eds), *Metalle der Macht – Frühes Gold und Silber. Metals of Power – Early Gold and Silver, 6. Mitteldeutscher Archäologentag, 17-19. Oktober 2013*, Halle, 2014, 183-207 [*Tagungen des Landesmuseum für Vorgeschichte Halle* 12].

Z.A. Stos-Gale and C.F. Macdonald, "Sources of metal and trade in the Bronze Age", in: N.H. Gale (ed.), *Bronze Age Trade in the Mediterranean, Papers Presented at the Conference held at Rewley House, Oxford, December 1989*, Jonsered, 1991, 249-284.

P.G. Themelis, "Εργαστήριο χρυσοχοΐας του 8ου αἱ. π.Χ. στην Ερέτρια", *AAA* 14, 1 (1981), 185-208.

P.G. Themelis, "An 8th century goldsmith's workshop at Eretria", in: R. Hägg (ed.), *The Greek Renaissance of the Eighth Century B.C.: Tradition and Innovation, Proceedings of the Second International Symposium at the Swedish Institute in Athens, 1-5 June 1981*, Stockholm, 1983, 157-165.

M. Tiverios, "The ancient settlement in the Anchialos-Sindos double trapeza. Seven years (1990-1996) of archaeological research", in: M. Bats and B. D'Agostino (eds), *Euboica: L'Eubea e la presenza euboica in Calcidica e in Occidente, Atti del Convegno Internazionale di Napoli, 13-16 Novembre 1996*, Naples, 1998, 243-253.

M. Tiverios, "Greek colonisation of the northern Aegean", in: G.R. Tsetskhladze (ed.), *Greek Colonisation: An Account of Greek Colonies and Other Settlements Overseas*, vol. 2, Leiden and Boston, 2008, 1-154.

M. Tiverios, "The presence of Euboeans in the north Helladic region and the myths of Heracles", *Studies in Ancient Art and Civilization* 17 (2013), 97-112.

S. Triantafyllou and S. Andreou, "Claiming social identities in the mortuary landscape of the Late Bronze Age communities of Macedonia", in: J.M.A. Murphy (ed.), *Death in Late Bronze Age Greece: Variations on a Theme*, Oxford, 2020, 171-197.

A. Vacek, "Al Mina and changing patterns of trade: The evidence from the eastern Mediterranean", in: X. Charalambidou and C. Morgan (eds), *Interpreting the Seventh Century BC: Tradition and Innovation, Proceedings of the International Colloquium Conference held at the British School at Athens, 9th-11th December 2011*, Oxford, 2017, 47-59.

P. Van Alfen and U. Wartenberg, "White gold and the beginnings of coinage: An introduction to the current state of research", in: P. Van Alfen, U. Wartenberg, W. Fischer-Bossert, H. Gitler, K. Konuk and C. C. Lorber (eds), *White Gold: Studies in Early Electrum Coinage*, New York and Jerusalem, 2020, 1-16.

M. Vavelidis and S. Andreou, "Gold and gold working in Late Bronze Age northern Greece", *Naturwissenschaften 95*, 4 (2008), 361-366.

M. Vaxevanopoulos, J. Blichert-Toft, G. Davis and F. Albarède, "New Findings of Ancient Greek Silver Sources", *JAS 137* (2022), https://doi.org/10.1016/j.jas.2021.105474.

S. Verdan, *Le Sanctuaire d'Apollon Daphnéphoros à l'époque géométrique*, Gollion, 2013 [*Eretria* XXII].

S. Verdan, "Metallurgical Ceramics from the Hypogeion", in: Morris and Papadopoulos 2023, vol. 1, 545-562

S. Verdan and E.D. Heymans, "Men and metals on the move: The case of Euboean gold", in: T. Cinquantaquattro and M. D'Acunto (eds), *Euboica II. Pithekoussai and Euboea between East and West, Proceedings of the Conference, Ischia, 14-17 May 2018*, Naples, 2020, 279-300 [*AION n.s.* 27.1].

E. Voutiras, "La réconciliation des *Dikaiopolites*: une nouvelle inscription de Dikaia de Thrace, colonie d'Érétrie", *CRAI*, 2008, 781-792.

K.G. Walker, *Archaic Eretria. A Political and Social History from the Earliest Times to 490 BC*, London, 2004.

K.A. Wardle, "Excavations at Assiros Toumba 1987: A preliminary report", *BSA 83* (1988), 375-387.

K.A. Wardle, "Excavations at Assiros Toumba 1988: A preliminary report", *BSA 84* (1989), 447-463.

K.A. Wardle and D. Wardle, "Assiros Toumba: Remains of the later Iron Age", in: P. Adam-Veleni (ed.), *Μύρτος: μνήμη Ιουλίας Βοκοτοπούλου*, Thessaloniki, 2000, 653-673.

6. ORE PROCESSING TECHNIQUES IN THASOS DURING ANTIQUITY

Sabine Nodin

Introduction

Ore processing is a crucial stage in any metal production workflow, which comprises all the operations converting natural minerals extracted in their raw form into an enriched product, a processed material close to the desired mineral. The study of Greek mineral processing has traditionally been based on the Athenian example of Laurion, which actually represents a unique case of silver extraction on a large scale.[1] Although northern Greece, endowed with gold in addition to silver deposits, has been another important region contributing to ancient wealth, its history of ore processing has not attracted equal attention. While mineral deposits and – to a lesser extent – mining activities are well documented,[2] the mineral-processing activities have not been studied in detail.

Although the sophisticated model for the treatment of lead-silver ores in Attica is well-known, such evidence in lacking in northern Greece, in particular due to the absence of comparable enrichment facilities (edge-runner mills, washeries etc.). The reasons for this deficit are certainly related to the history of mining and to the limited targeted research, while it also depends on geological realities, mainly linked to the major difference in the North Aegean mineralisation compared to that of other regions, in particular its polymetallism. The Thracian environment is different, characterised by a higher hydrometry and a much denser forest cover. From this point of view, it is legitimate to question the diversity of the technological processes concerning post-mining processing. Crushing and grinding are necessary operations for the metal production *chaîne opératoire*, but could it be that each phase corresponds to a stable typology of tools devoted to a single function? It has been argued that during the various periods of Greek antiquity ore-processing tools are virtually indistinguishable from those used in cereal processing for flour making. This was therefore the essential research question of the current study, to reassess the functional hypotheses related to these tools, in the light of a specific archaeological context. Additionally, to determine whether there is specialization and to be able to identify the different functional bridges between various supports and to consider how the diversification of this toolset compensated for the lack of "labour specialisation" in ore-processing activities. Careful examination of previous studies on each of the mineral-processing phases demonstrates a complete gap in our understanding of the ways by which this toolset is intertwined with its associated *chaînes opératoires*.

This article is based on a Masters thesis entitled "Lithic Tools Employed in Greek Ore-Processing Based on the Example of Thasos" that attempted to fill this gap in research at an international level, aiming to explore the reconstruction of techniques in order to be able to envisage the organization of production in a diachronic way. Examination of 259 unpublished lithic tools forming the corpus of this study allows a considerable progress in our knowledge on the various ore processing *chaînes opératoires*, which are not clear so far. The way in which these processes unfolded is crucial for a better understanding of the subsequent stages of smelting and production of metals.

The ore crushing technique Mineralurgical *chaîne opératoire*

The rationale for the study of stone tools and the necessary adaptations from related disciplines to address archaeological materials have underlined the need for a multidisciplinary approach, quite challenging in the sense that the analysis stands at the intersection of geology (petrography), traceology, archaeology and the history of techniques (extractive, mineralurgical[3] and metallurgical). The current examination of the material forming this corpus tends to verify the possibility of establishing criteria of discrimination between mineralurgical tools and

1 Ardaillon 1897; Conophagos 1980; Papadimitriou 2017.

2 Wagner and Weisgerber 1988.

3 The term "mineralurgy" derives from the French terminology and refers to the treatment/processing of mineral ores.

grinding instruments. This allows the combination of each of these approaches for a holistic study of the tools, to bring about considerable progress in our knowledge of the various mineralurgical operating chains. Although Laurion represents a well-known example for the processing of lead-silver ores, when examining the north Aegean we face several methodological problems, in particular the absence of traces of comparable enrichment facilities. Thus, by combining data that have not been considered in an overall perspective, we attempted to extract information allowing us to have a somewhat clearer idea on a diachronic level about the different trituration processes employed on mineral ores.

A quick perspective on metal deposits allows for an initial step in approaching the issue. Metalliferous ores come from natural mineral formations containing various metals, very rarely in their native state. These fall into two major groups: monometallic and polymetallic. In order not to confuse the notions of "mineral" and "ore", it is then necessary to specify the notion of "ore", that is to say a natural substance composed of both useful and useless minerals. We can take the example of iron ore: it consists of useful minerals i.e. hematite, limonite and useless minerals such as quartz. All minerals are classified according to their structure into two groups: disseminated and non-disseminated. It is still possible to subdivide the disseminated ores into three groups: the fine dissemination having inclusions or mineral grains of dimensions less than 40-50 microns, the medium dissemination having grains of 120 to 200 microns and the coarse dissemination whereby grains can go up to a few millimetres. For disseminated ores, the main role of the preparation processes is to liberate the grains, otherwise enrichment would be impossible. Thus the term ore is generally applied to rocks containing metalliferous minerals in their native state (gold, silver, copper, etc.) or combined with other elements to form more complex minerals: oxides (iron, etc.), sulphides (lead, zinc, etc.), silicates (nickel, etc.), carbonates, phosphates and other types. The metal resources on the island of Thasos[4] are divided into three categories: ferrous metals often containing impurities (chromium, nickel etc.), non-ferrous metals (antimony, arsenic, copper, lead, zinc etc.) and precious metals (gold, silver), the latter being present in very low quantities. According to their chemical composition, metallic minerals are defined as:

- native (gold, silver, copper) or primary;
- oxidized, or secondary, i.e. having undergone an alteration subsequent to their deposition (for example pyrite, a primary mineral, becomes goethite, a secondary mineral; chalcopyrite, a primary mineral, is transformed into azurite and malachite, secondary minerals); oxides and hydroxides (iron: goethite FeO(OH), hematite Fe2O3, limonite $FeO(OH) \cdot nH_2O$ (variety of goethite), magnetite Fe3O4; hydroxyl carbonates of copper: malachite Cu2CO3(OH)2, azurite Cu3CO3(OH)2); carbonates (cérusite CaC03); sulphides (anglesite PbS04);
- sulphidic (galena PbS, blende ZnS).

This preliminary work makes it possible to approach the different archaeometallurgical processes, the present part of which will endeavor to distinguish the sequences in order to approach them within their own technical system. Due to the very nature of the anisotropic materials present on the island, which highlight a great heterogeneity due to the irregular distribution of their constituents, it is reasonable to admit that the materials subjected to the last comminution operations (crushing, grinding) may give rise to different techniques:

- for oxidized minerals such as iron ores, these would simply be subjected to reduction operations following the bloomery process [5];
- the sulphide minerals would have to undergo preliminary heat treatments in an oxidizing atmosphere, mainly roasting to drive off the sulphur and transforming them into oxides. This treatment would also be applied to the carbonates, with carbon being converted to carbon dioxide. Finally, the hydroxides can also be freed from their water, allowing the iron to combine with oxygen in the air to form oxides.

Studies of ancient mineral processing require a thorough interpretation and clear understanding of the *chaîne opératoire* in the regional context within which the various operations are embedded. In order to be able to deepen our knowledge, it is therefore essential to have adequate elements of comparison. The insufficient study of this toolset, the use of which, in this context, remains generalized and imprecise, has so far been limited to an intangible succession of tools required for the preparation of the blocks of rock extracted until their transformation to metal. However, reconstructing these sequences might be challenging since they often leave minimal traces in the archaeological record.

4 Vavelidis *et al.* 1988, Bassiakos 2012.

5 Mangin 2004, p. 8.

The analysis of the different parts of the corpus will thus constitute a first assessment in order to specify these different operational chains and to propose a plausible scenario outline. This work must serve as an example for a thorough study with a good understanding of the archaeological remains, making it possible, through a rigorous methodology, to optimize the future experimental achievements essential to the creation of a corpus of a use-wear reference inventory. This would in turn make it possible to accumulate relevant information in order to be able to consider assessments on the island as a whole and to validate the archaeological interpretations. Such a deductive interpretation, induced by the study of the actual material and by the intervention of knowledge from several fields of research, will revolve around the production chains of copper, iron and argentiferous lead associated with the different Thasian activity areas.

Our goal is to reconstruct the working processes of materials, to understand the choice of rocks, the evolution of the forms of tools, whether specialized or not, as well as the beneficiation techniques implemented in antiquity in order to render metallurgy more effective. It is therefore important to remember the following principles specific to ore crushing, which can be divided into three to five phases:

- The first corresponds to their extraction: once the ore was extracted from the mine, it was subjected to a second manual sorting, which completed the initial choice undertaken within the mine. Any chunks that were less than 10% or 12% rich in the desired ore were considered sterile and the rest was classified into two categories. Pieces between 12% and 30% rich were treated further while those above 30% were ready for direct smelting.
- Comminution corresponds to the second phase in mineral processing where the extracted rock is successively reduced by fragmentation until the necessary particle size is obtained to separate the useful mineral grains from the waste rock. This particle fragmentation occurs through crushing (compressive failure), impact (tensile failure), wear/grinding or attrition (shear failure). The terminology used to designate the operations performed depends on the operational phase, distinguishing however two main primary mechanisms:
  - Crushing is the first mechanical step in the comminution process. This is a typically dry process with small reduction ratios. Its first Phase is referred to as "primary crushing", mainly aimed at reducing the size of the rock by compression or impact in order to make it suitable for its future application. In impact fragmentation, the particle is suddenly exposed to very high stresses where the particles break by pulling into fragments of similar size. This fragmentation involves several successive steps, each of which reaches an increasing level of fineness.
  - Grinding takes place mainly by compression and friction between two solid surfaces. The tensile stress is generated by the simultaneously perpendicular and tangential displacement of the two active surfaces with respect to each other. The compression induces particles of two types of sizes: the fine ones from the compressive failure, which occurs near the load points, while the coarse particles occur following the induced tensile failure. Unlike crushing, grinding is a typically wet process, both to ensure a degree of cohesion of the material and to reduce wear on the grinders made of volcanic rocks. Wet comminution offers some advantages, since water has a high surface tension and a very low viscosity.[6] The internal tensions are never uniform because the rock contains many different minerals whose sizes are variable. While the stresses depend on the mechanical properties of each constituent as well as the presence of cracks, water reduces the energy required for fragmentation by lowering the bond strength at the tip of the cracks. Also in this way, the dust does not produce clouds of airborne ore, a feature that is valuable when the crushed material is particularly hazardous to the health of workers. Finally, by using ridged surfaces as can be seen on some grinders, the surface area of the load point will reduce the compressive forces in order to separate the constituents already released and therefore avoid "overgrinding", which generates losses during flotation. However, the finer the ore is ground, the better the targeted release of the ore will be. The particle size reduction ratio will therefore depend on the types of mills and the material being processed.
- Washing is the third stage of mineral enrichment, using gravimetric reasoning i.e. taking advantage of the difference in specific weight of the mineralogical constituents where particles are separated by physical means based on their properties such as density.
  In the case of sulphide ores, they will have to undergo a more demanding treatment requiring

6 Lowrison 1963, p. 103.

two additional steps: roasting (sometimes repeated several times) and reduction strictly speaking (slag fusion). Indeed, for the secondary ores such as oxides and carbonates these could be reduced in one step at relatively low temperatures[7] and therefore do not require these treatments. Previous investigations have shown that the intensive exploitation of secondary mineralized layers, has forced ancient metallurgists to develop new processes when they had to reach the sulphide ores of the primary layer, which were much more difficult to process.

- Thus, once the rock has been reduced to a sufficient size by crushing, it is usually roasted. The roasting operation is essential for the smooth running of subsequent operations. Roasting is the thermal operation consisting of the elimination of certain chemical bodies, such as sulphides and phosphorus compounds that are unwanted in metallurgy. This thus makes it possible to facilitate the subsequent grinding step by oxidizing the sulphides, eliminating sulphur and arsenic and weakening the rock which has become more friable by releasing the grains caught in the sulphide networks.
- After washing, the ore was loaded into a reduction furnace with fuel and, if necessary, additives (fluxes) to aid the operation. Smelting is a decisive process, which aims to transform the gangue into slag readily separable from metal by density. This time, high temperatures must be used to allow the metal to flow through the slag. In the case of reducing sulphides without significant quantities of matte being produced this implies that the roasting stage was sufficiently effective. Otherwise, any produced matte would have to undergo secondary crushing in order to recover quantities of metal entrapped therein.

The question of Ore-processing / metallurgy in Thasos

The principles described above allow for representative reconstructions of processes whose specific investigation helps us to better understand the development of metallurgical activities in their evolutionary phases. From the basis set by our previous analyses, we will now attempt to examine and compare the results related to exploration and mining as well as metallurgical evidence on Thasos, cross-referenced with ancient sources, in particular at Limenas after the Parian colonization (Fig. 1).

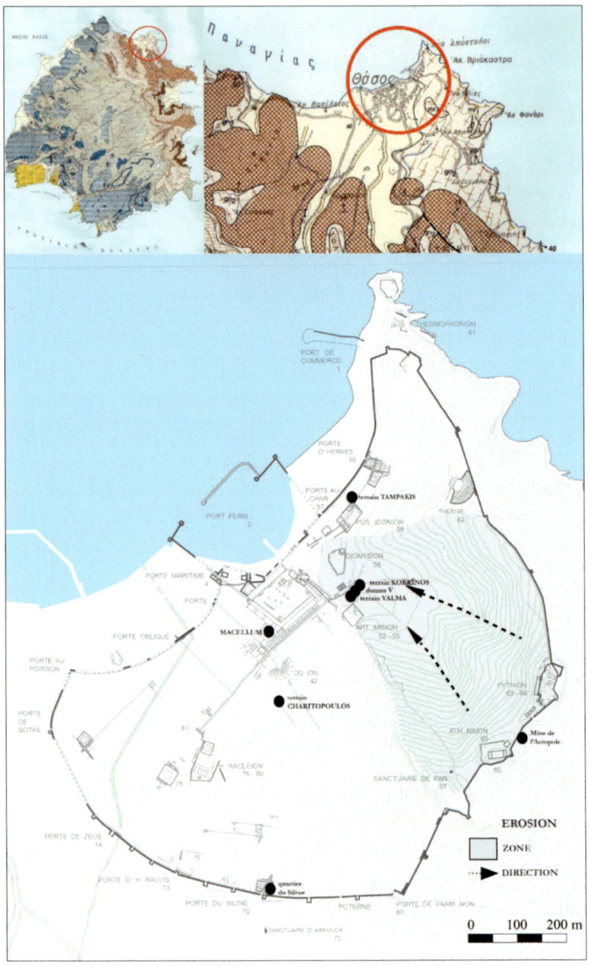

Fig. 1. Limenas indicated on the geological map of Thasos (source: IGME) and plan of the city. Black dots indicate metalworking contexts (source: EfA/M. Wurch-Kozelj and T. Kozelj).

Although limited, the data extracted from the various excavations or soundings are based on a chronology deriving from well-defined contexts with a clear stratigraphy. These can be used to allow for a synthetic approach, from the mine of the acropolis and the metallurgical workshops located downstream that were in direct relation with the various establishments. This approach will be carried out with regards to the situation on the whole island as well as its Balkan hinterland, from which it has only been separated since the Neolithic period.[8]

7 Budd *et al.* 1991 demonstrated that a temperature of 700°C was sufficient.

8 The island of Thasos was part of the eastern Macedonian mainland, without a maritime channel separating it from the opposite coast in the Upper Quaternary, until 8000-5000 BC. see Lambeck 1996; Perissoratis, Mitropoulos 1989.

The metalworking context at Charitopoulos plot, Limenas

The first site of interest is located southwest of the Odeon, at Charitopoulos plot, that was partially investigated in 1985 before being excavated in 2008 and 2009 by the Ephorate of Antiquities of Kavala. The material studied here comes from a private building (complex II) extending on both sides of the Charites Gate and the earliest Phase of the Archaic fortification. Remains of a workshop have been unearthed, featuring a crucible-furnace (K4-K8), an associated channel, and a well (K6) conventionally dated to the first half of the 6th century BC (570 BC at

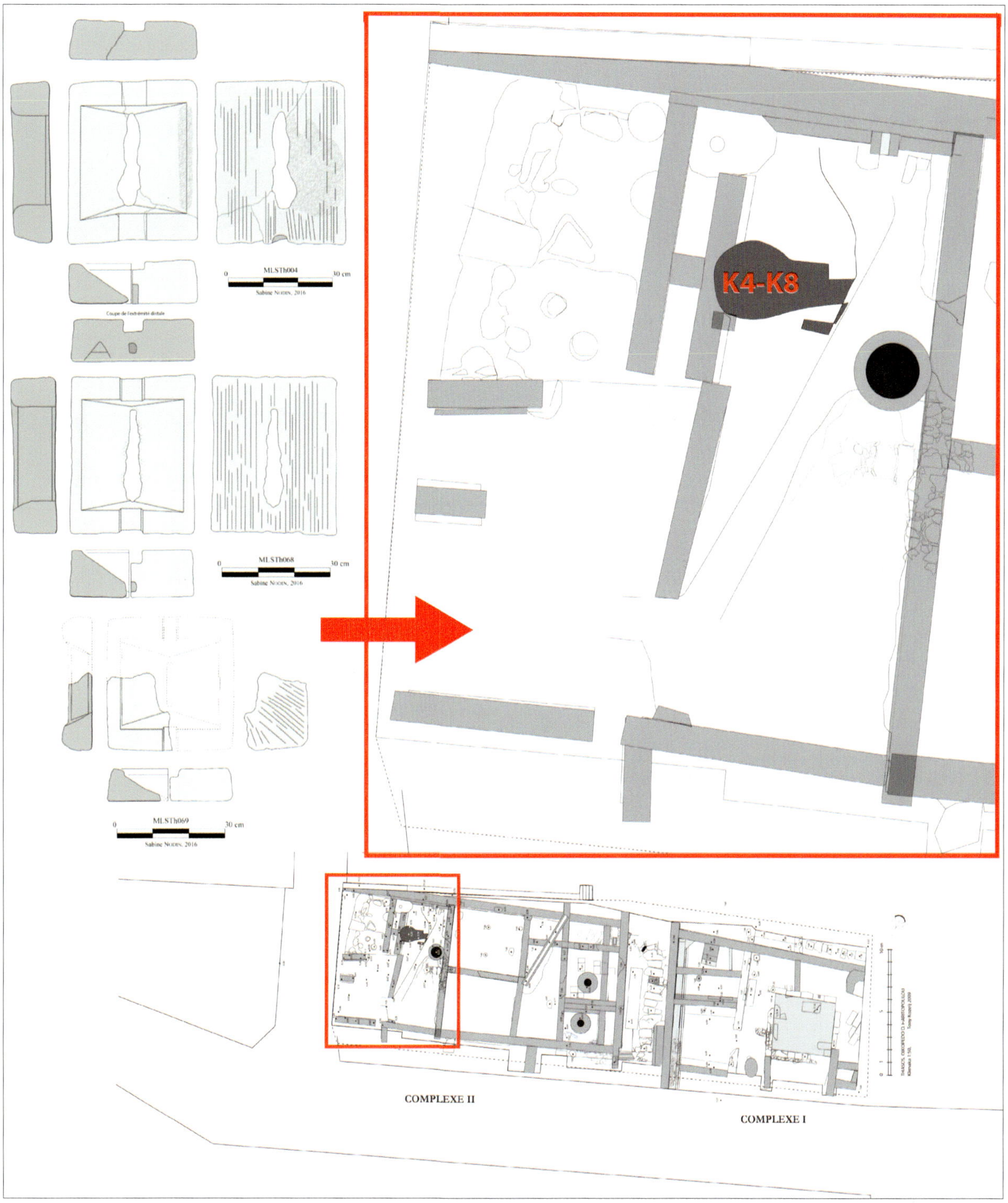

Fig. 2. Charitopoulos plot with a focus on complex II (plan: T. Kozelj). The arrow indicates the location of the three Olynthus type millstones (drawings: S. Nodin 2016).

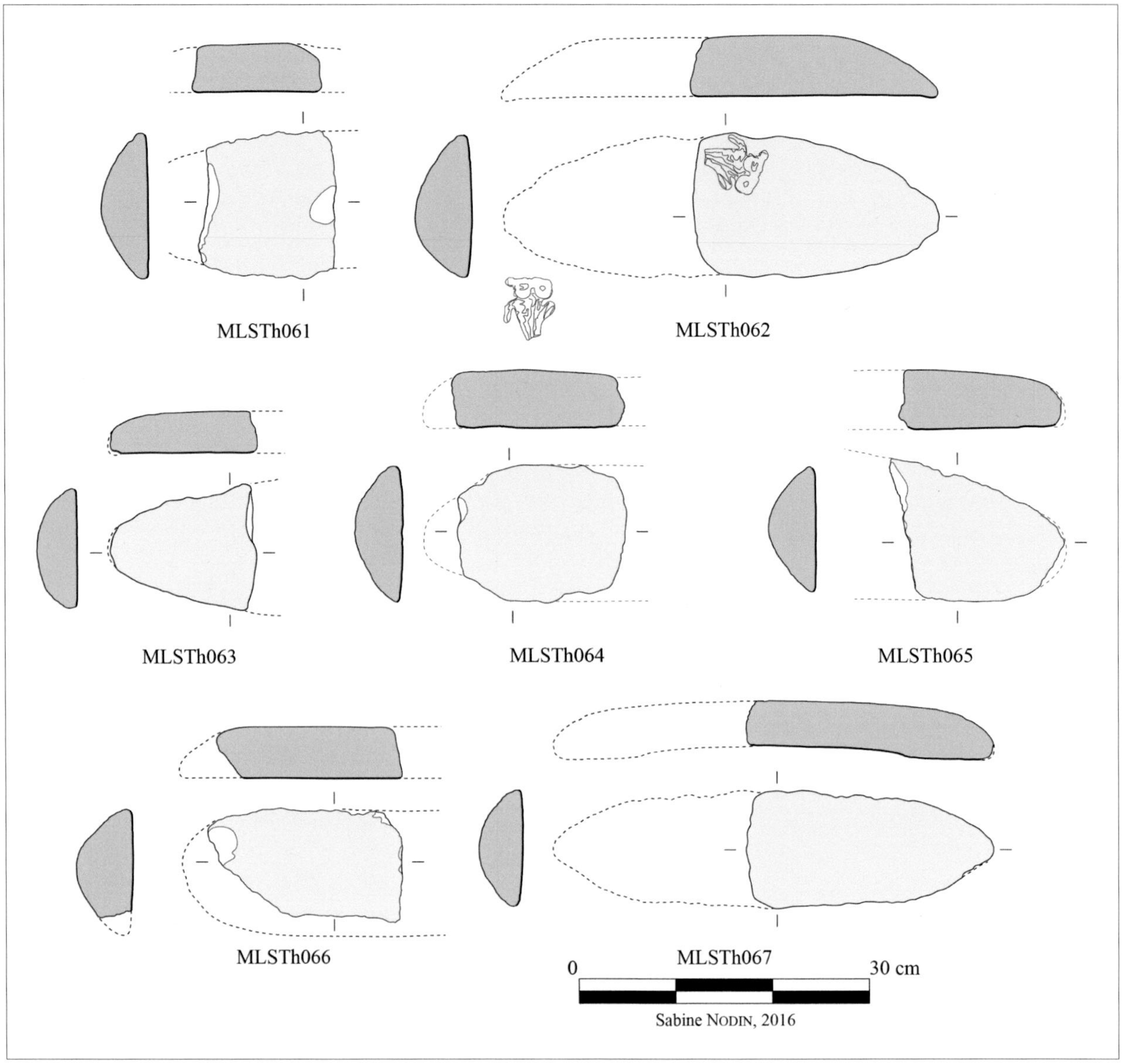

Fig. 3. Millstones of saddle quern type found during the 2008-2009 excavations at Charitopoulos plot, Limenas.

the earliest) but whose backfill could be dated from the beginning of the 5th century BC at the latest (Fig. 2). During the first excavation of 1985 four saddle-querns were found, three mobile (040-041-042) and one dormant (055)[9]. Ten millstones, all mobile were unearthed during the second excavation campaign: three of the Olynthus type found in the room located to the west of the workshop, housing the metallurgical installation and seven saddle-querns (Fig. 3). It should be noted that two dormant millstones associated with MLSTh004 and MLSTh068 present on the photographs taken during the excavations were not found in the museum's storerooms and could not be included in the framework of the current research. As regards the K4-K8 furnace, its embankment could be dated to the beginning of the 5th century BC. The metallurgical activities are therefore subsequent to the earliest occupation of the site.

The analysis of the active surfaces of the millstones has made it possible to highlight different uses: it thus appears that two of these tools have residues resembling iron oxides (**004**, **068**), while the third seems to bear residues of lead ores (**062**). One millstone displays traces of multiple uses, with concomitant residues of carbonates of copper, iron

9 We are entitled to wonder about the specific methods of abandonment, deposition or rejection of these millstones that were undoubtedly deliberate. One might ask why were these three unused "new" tools found here and not necessarily adapted to the dormant table that accompanied them.

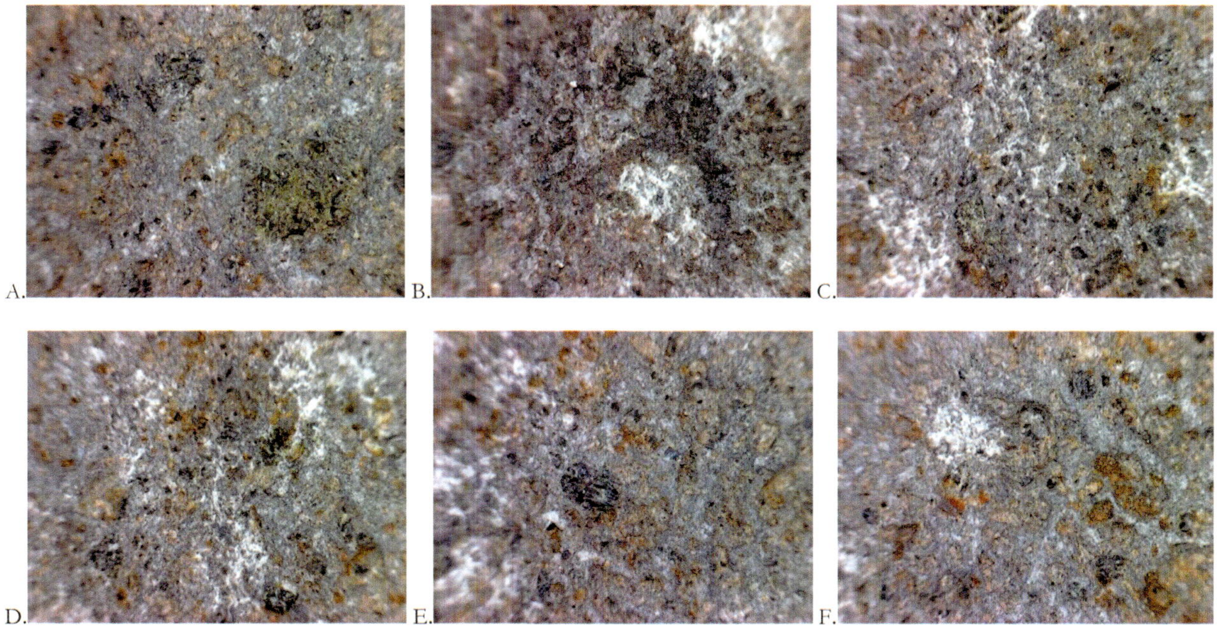

Fig. 4. Traces and residues on the surface of millstone 064 - saddle quern type. The traces observed on its active surface seem to bear witness to multiple uses for this tool, namely in two functions both as a grinder and/or a percussion tool. It displays densely struck edges and a surface worn out by strikes of launched percussion concentrated at the center, associated with residues of carbonized material (e.g. D. and E. show residues of charcoal). Stigmas of dense distribution are accompanied by worn grains, a blunting appearance parallel to microfractures of alteration whose characterization allows the distinction of micro-polish and micro-impacts as well as larger fractures linked to grain tearing. These traces reflect the same processing chain but testify to a versatile context meeting the needs of a polymetallic activity. Notable co-existence of greenish spots of copper carbonates, which could be malachite (A., D. and F.), scattered traces of punctiform spots of reddish to yellowish colour indicating a probable work of iron ore (F.), fragments of greenish-yellow colour, possibly a lead containing mineral such as cerussite (D. and F.) and residues of silver-white, lustrous material, associated with deposits of dark and relatively shiny ore similar to lead (?). It also seems that this grinder was used for grinding after roasting (?). Magnification x40 (Nodin 2016, 121).

and lead (**064**) (Fig. 4). The traceological analysis could validate the grinding of hard mineral matter for three more tools whose examination gave no visible residues, but whose stigmata found scattered on their active surfaces attest to an intense abrasive action (**061**, **066**, **067**). Thus, all these elements testify to polyvalent contexts meeting the needs of an intense polymetallic activity, possibly associated to the direct exploitation of the mine on the acropolis located some 200 m uphill, (the presence of a canal dug in the same sector would also have facilitated the washing of the ores during the enrichment phase), but also depending on supplies of sulphide ores, perhaps brought from the western part of the island. Concerning the latter (namely galena and / or blende and / or chalcopyrite), if we could establish the nature of the copper ore (oxidized, sulphide) by studying slag inclusions, we could hypothesize that chalcopyrite was first crushed and then roasted, perhaps directly in the K4-K8 oven, whose size and shape do not exclude such use, then hammered and ground before the smelting phase. So far, the metallurgical evidence from this site does not point to any copper smelting activities, rather the melting and recycling of bronze[10]. In the same way for the traces of lead sulphide, we do not currently have any metallurgical evidence at this exact location, which does not mean that it did not exist, somewhere in Limenas or near the mine of the acropolis (tool 053 points in the same direction, although the origin of this crusher is unknown), without doubt in plots that have not been excavated or among material that has not been studied so far. The director of Speidel Mining Corporation, J. Speidel indicates as early as 1929 the presence of numerous metallurgical residues (mattes and slag) on the south-west slope below the entrance to the mine, which might suggest smelting activities in this sector.[11]

The second important point concerning this context is that excavation has also brought to light a unique object of which we have found no parallels in the

10 Nerantzis *et al.* 2017.

11 Speidel 1929.

5
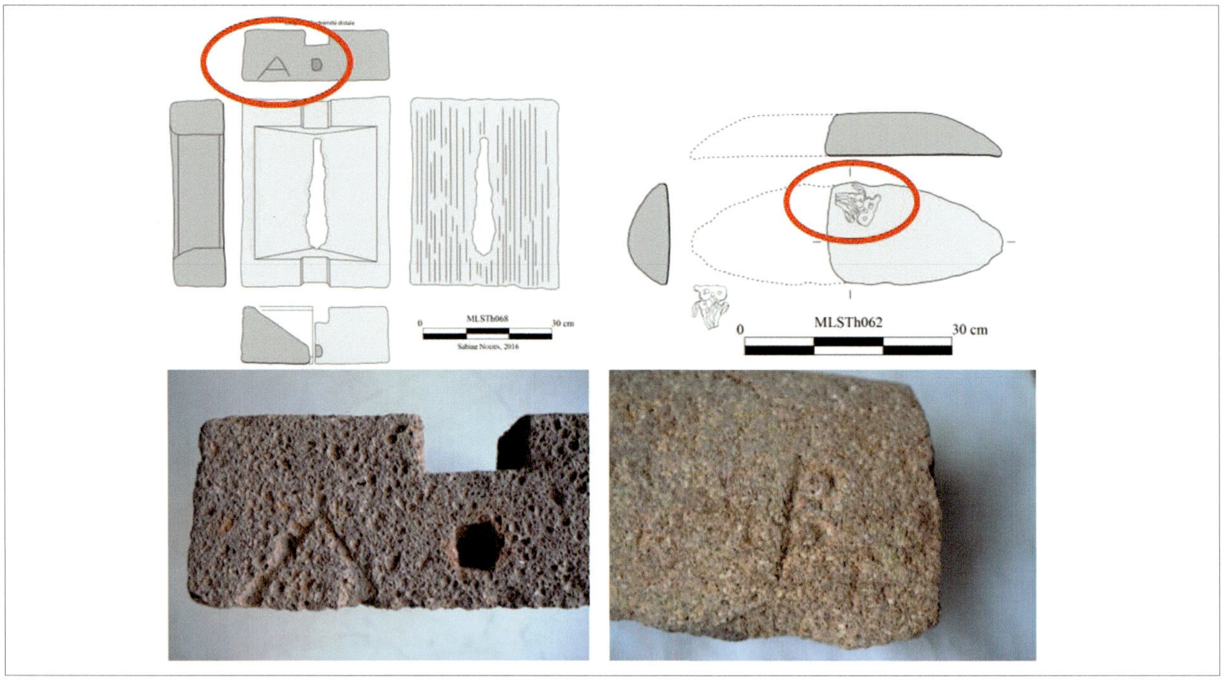

6
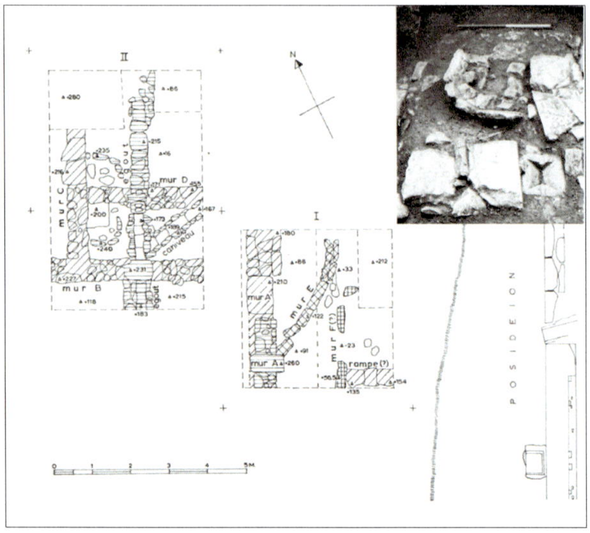

Fig. 5. Olynthus type millstone marked with A (left) and saddle quern marked with B (right) from Charitopoulos plot. The context at Tampakis plot.

Fig. 6: Tampakis plot, location plan and overall survey of the excavation; photo of the South part of the sewer with MLSTh052, seen from the East (Maffre 1970, fig.43 and Fig. 63 combined)

relevant literature. The distal end of mill no. **068** made of porous basaltic andesite is marked with the letter "alpha" (Fig. 5). Another grinder in the corpus, unfortunately of unknown provenance, has exactly the same mark on the same edge (**054**). Associated with this first hopper, a small saddle quern no. 062 made of massive dacite has its proximal edge marked with a "beta" (Fig. 5). This raises questions concerning an evolving type pattern in the succession of operations of the *chaîne opératoire*, a hypothesis corroborated by the presence of residues from the metallurgical phase itself on grinders **062** and **064**, while on the other hand revives the question of sizing and of a succession of granulometric order. From this point of view, we may be surprised that the excavation reports have not mentioned the proximity of saddle querns, generally considered to be Archaic, and Olynthus type millstones, considered to be dating to the Classical and Hellenistic periods. Broadly speaking the former are considered to last until the latter gradually replaced them, yet as we can see very clearly from this study, this should not be the case. Could each of these tool types have been used for different purposes? This is what we are entitled to think from the initial results obtained. In the case of Charitopoulos plot, the Olynthus type, and perhaps the three saddle querns **061**, **066**, **067** could have been used to fragment the ore before roasting or direct smelting. For those who had to be roasted, upon completion of the roasting stage workers would have used an old saddle quern, **064** as a hand hammer this time, before finely grinding the ores using **062**, including the "beta" tool in a secondary phase of use.

The context at Tampakis plot

Two butterfly-shaped hopper Olynthus millstones (050-052) were found during rescue excavations at Tampakis plot, located a few meters northwest of the *Poseidonion*, practically in the main alley of the sanctuary, in layers dating from the Hellenistic to the Roman period (figure 6). The layout is fairly complex consisting of a small sewer sloping slightly from SW to NE, a small open-air gutter oriented EW and communicating with the sewer and a semi-circular stone structure surrounding a stone paved area (which would undoubtedly be a basin). This layout has been built sometime in the 2nd century BC and was used continuously until the 4th century AD. One of the millstones was lying on the ground, the other had been reused as a slab for the sewer. According to Maffre, this set may have been an enclosure forming part of an agricultural or industrial installation.[12] Examination of the first grinder revealed traces of processing lead-bearing minerals (**050**) and the second had residues of lead and iron-rich minerals (**052**). The study of the only crusher of the same type (**051**) found in the storerooms of the museum of Limenas, of unknown provenance, also indicates processing of lead sulphides such as galena associated this time with the elaboration of flux (gold bearing quartz?). Coincidence or not, the same type seems to be used for the same purpose.

Macellum

The treatment of objects from recent excavation campaigns on the southern outskirts of the Hellenistic Agora of Thasos (1996-1998 and 2009-2010) made it possible to identify two parts of the *macellum* dating from the 3rd century AD that may have been used for metallurgical craft activities: room 54 was devoted to bronze working as attested by the finding of slag and other metallurgical waste associated with fragments of moulds, crucibles, nozzles and furnace wall fragments and at least four grinders of the Olynthus type (figure 7). Room 57 would have been devoted to ironworking, based on relevant finds but accurate dating of the context is lacking while it was not possible to specify at which stage(s) of the operational chain this waste belongs to. This room yielded two grinders, a fragment of the Olynthus type (**109**) and a fragment of the gyratory type (**319**). The elements studied seem to confirm the hypothesis of copper working in room 54. However, one element attracts our attention: the presence of a "circular structure" in the room's interior. Even though we lack data on the origins of the 12

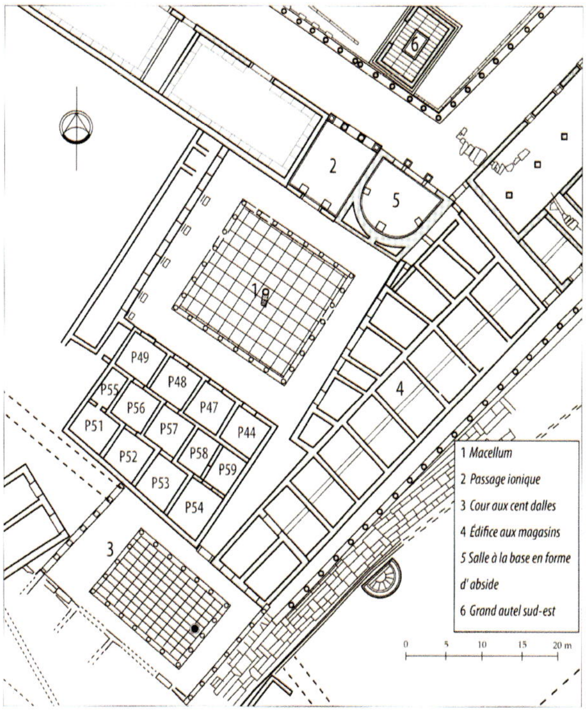

Fig. 7. Plan of the *macellum* and its surroundings. Marc 2011, Fig. 1.

crushers resulting from this excavation, we still note the presence of 2 gyratory crushers, perhaps to be put in relation with another mobile element of the same type, **044** made from exactly the same rock, found in the agora in September 2004, and whose handling has made it possible to suggest assembly hypotheses. The whole cone crusher **057** was also found in the Hellenistic Agora, without further information on its background. Could the circular structure mentioned in exhibit 54 be a gyratory millstone commonly referred to in relevant literature as a "delian type"? Traceological examination of the active surface of **319** affirms grinding of hard mineral matter, the residue examination of which perhaps suggests the grinding of quartz or calcite necessary in the preparation of fluxes that the two workshops might have needed.

Variability and typological evolutions, technological transfers

With regards to the periods under study, we have been able to note the joint presence of two types of millstones within various contexts which undoubtedly relate to installations where ore enrichment may have taken place (figure 8). Since the Archaic period when alternative manual types could have been associated with lever grinders, they have evolved until the Roman period where this time the latter accompany gyratory grinders, which demonstrates an evolution in grinding

12 Maffre 1970, 844.

techniques. We notice that if the saddle quern type was cut in exclusively massive rocks, the intermediate Olynthus type, of large dimensions made from volcanic rocks characterized by large pores, seems to undergo a morphological change, becoming on the one hand smaller, compact, with a selection of rocks that are still porous, but whose pores are much finer in size. To this first change we should perhaps consider a second, one of a technological nature this time. It was hypothesized that the large grinders from Charitopoulos plot could have been used as a preamble to fragmentation actions, directly linked to the enrichment of the ore. This is why some were labelled with an "alpha". Reading the marks imprinted on the active surfaces shows different scales of intensity evolving from abrasive crushing to refining from the mesial part to their edges and ends. This crushing around the millstone slot could be described as intermediate, with primary crushing probably occurring at the mine's vicinity. It is interesting to note that no pestle, hammer or anvil was found near the grinders used for the fragmentation of the ores and we can at most hypothesize that these *ergasteria* perhaps worked with an ore "ready to use"[13]. The millstones of the saddle quern type, like **062** marked with a "beta" then allowed a second finer grinding until reaching the desired particle size. If the tool was hardly damaged with use, it could also be reused as a hand hammer such as **064**, in the same way that the broken hoppers would be reused as hammers. During the Roman period and despite the fact that manual rotating grindstones spread throughout the Greek world, one observes in Thasos a conservatism in using "Greek" type grinders. And this is where there is technological transfer, in the sense that the lever grinders, the proportions of which tend to be gradually reduced, still with porous rocks but being selectively more compact, with smaller pores, adjoin gyratory type cone grinders. Apparently, they are still in use, but they seem to have replaced the saddle quern. For crushing and primary grinding, we could hypothesize that the massive Olynthus type could have been replaced by gyratory grinders which would have made it possible to improve the quality of enrichment, thus representing an innovation perfecting the grinding and increasing the volume of materials that it became possible to process continuously. However, this hypothesis is based on limited evidence of flux grinding indices and will need to be further explored.

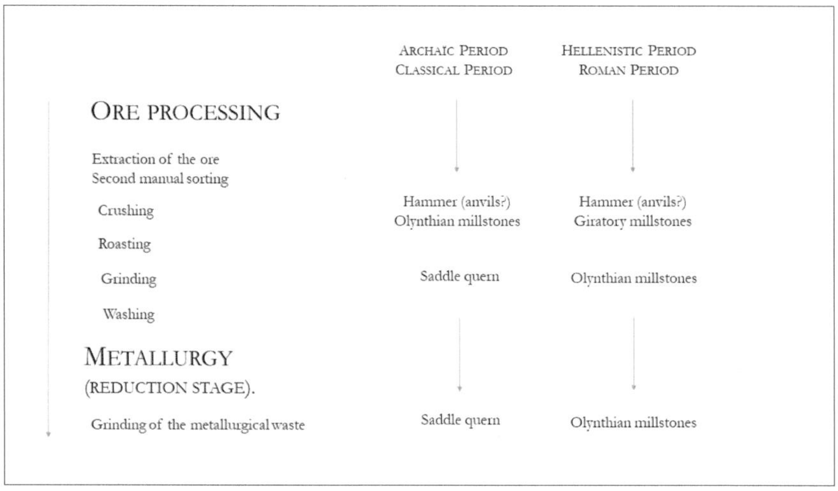

Fig. 8. Proposed model for the Thasian toolsets embedded in the mineralurgical and metallurgical chaîne opératoire.

Co-presence of hydraulic installations and grinders

Most of the findings presented come from private facilities, used diachronically and yet it seems that none of these facilities have been thoroughly investigated so far. At Charitopoulos plot, a well with a diameter of 0.80 m and a stone drain were partially excavated. Due to the opening of a trench during WW II by the Bulgarian army, the drain and associated layers suffered serious disturbance. However, this

13 We also reflected in this respect on the impact of metal on lithic technology, namely, at what period could a transfer have taken place from tools made of stone to those made of iron. According to Agarthachides of Cnidus, iron pestles were used in the gold mines of Egypt in the 2nd c. AD See Photius, Library VII, Codex 250, Agatharchides, 26, trans. Henry R., 1974: "(...) Those then who are less than thirty years old and whose blows are vigorous, take stone mortars and, with iron pestles, they crush the stone vigorously and break it into pieces. the largest of which is roughly like a lentil". At Thasos, the eastern slope of the mountain called Fanos or Klisidhi conceals many vestiges of intense mining, three sectors of which have been the subject of more precise research, and have yielded, among the material collected, an iron mass of 13 x 9.6 cm (see Des Courtils *et al.* 1981 and 1982, p. 410 and Fig. 3 p. 411); the authors specify that the state of preservation does not allow us to assert that its shape and its contours are original, even if the hypothesis seems plausible to them. At the same time, many stone tools can still be found in the bronze smith's workshop in room 54 of the macellum of Thasos, dating to the 3rd century. AD.

structure tells us about the spatial organization of the workshop. The finds at Tambakis plot could allude to the presence of facilities also devoted to this use in the Poseidonion area, which seem to have been functional during the final Phase of habitation between the 2nd century BC and the 4th century AD. The grinders on which galena processing residues were found could have been used in the same phase with the hydraulic installations. This reflection on enrichment contexts, which are seemingly not far from metallurgical contexts, brings together the question of production contexts and leads to a series of new questions based on purely hypothetical parameters but which may be of interest. We know that the mine of the acropolis did not yield any galena, but nevertheless Thasos was in need of silver for the minting of coins and one can thus reflect on the problems of controlling resources and production areas which must have played an important role very early on in Northern Greece. Although it is premature to determine if mining was under direct state control or pursued by private investors, one can at least suppose a centralized indirect control; the mines were perhaps guarded and likewise, the enrichment and metal production workshops had to be under close administrative control.[14] The Poseidonion area is interesting in this respect in the sense that it has always been part of the enclosure, whether in the Archaic period when it was also not very far from the so-called agora of the Charites, as after the extension of the rampart, where it remains relatively close to the Hellenistic agora.

Conclusion

The current examination of lithic tools comprising the Thasian corpus helps to establish discriminating criteria between ore-processing tools and milling instruments used in flour production. It has also highlighted the need for a multidisciplinary approach combining an analysis at the junction of geology, traceology, archaeology and the history of techniques (extractive, ore processing and metallurgy). Through the synthesis of acquired data that had not been considered before under a holistic perspective, we were able to extract information allowing us to have a slightly clearer idea, on a diachronic level, of the different processes of ore crushing and grinding in ancient Greece. We are however aware of the incomplete and non-exhaustive nature of this study, which nevertheless constitutes a first step towards a new field of research in order to be able to expand our knowledge on the various ore-processing operating chains. The lack of comparative evidence which we encountered and in particular the absence of enrichment facilities comparable to those at Laurion underline the fact that the lithic material should be taken into consideration on its own right. The island of Thasos was once located at the crossroads of maritime trade routes and its history demonstrates a favourable context for cultural exchanges conducive to technical innovations. While the question of technical exchanges with the indigenous Thracian population has remained unanswered, the presence of a wide variety of advanced crushers underlines the dynamism of the island's economic and social life.

14 Nerantzis 2015, p. 75-76.

Bibliography

É. Ardaillon, Les mines du Laurion dans l'antiquité, *BEFAR* 77 (1897).

Y. Bassiakos, "Πρώιμη παραγωγή χαλκού καί αργύρου στα Λιμενάρια Θάσου: μια τεχνολογική προσέγγιση", in: S. Papadopoulos and D. Malamidou (eds), Πρακτικά Ημερίδας Δέκα Χρόνια Ανασκαφική Έρευνα στον Προϊστορικό Οικισμό Λιμεναρίων Θάσου, ΥΠΠΟΤ-ΙΗ' ΕΠΚΑ, Θεσσαλονίκη, 2012, 197-224.

P. Budd, D. Gale, A.M. Pollard, R.G. Thomas, P.A. Williams, "The Early Development of Metallurgy in the British Isles", *Antiquity* 66 (1992), 677-686.

C.E. Conophagos, *Le Laurium antique et la technique grecque de la production de l'argent*, Ekdotike Hellados, Athens, 1980.

J. des Courtils, T. Kozelj, A. Muller, " Thasos : Mines antiques", *BCH* 105 (1981), 674-675.

J. des Courtils, T. Kozelj, A. Muller, "Des mines d'or à Thasos", *BCH* 106 (1982), 409-417.

A. Foucault and J.-F. Raoult, *Dictionnaire de géologie*[3], 1988.

K. Lambeck, " Sea-Level Change and Shore-line Evolution in Aegean Greece since Upper Palaeolithic Time", *Antiquity* 70 (1996), 588-611.

G. Lowrison, *Crushing and Grinding*, Butterworth, 1963.

J.J. Maffre, "Sondage Tampakis", *BCH* 94 (1970), 831-844.

M. Mangin (dir.), *Le fer, collection " Archéologiques"* (2004).

J.Y. MARC, "Les abords Sud de l'agora", *BCH* 135 (2011), 517-534.

A. MULLER, " La mine de l'acropole de Thasos", *Thasiaca, BCH Suppl.* V (1979), 315-334.

A. MULLER and T. KOZELJ, " La mine d'or de l'acropole de Thasos", in: G.A. WAGNER and G. WEISGERBER (eds) *Antike Edel- und Buntmetallgewinnung auf Thasos, Der Anschnitt*, Beiheft 6, Bochum: Bergbau-Museum, 1988, 180-197.

N. NERANTZIS, *Rhesos's Gold, Heracles's Iron: the Archaeology of Metals Exploration in Northeast Greece*, Potingair Press, Glasgow, 2015.

N. NERANTZIS, G. SANIDAS, B. JAGOU, T. KOZELJ and K. PANOUSSI, "An Archaic metallurgical workshop in Thasos (Greece): the case of Charitopoulos plot", *STAR: Science and Technology of Archaeological Research,* Vol.3, No.2, 2017, 148-160.

S. NODIN, *L'outillage lithique dans la minéralurgie grecque à partir de l'exemple de Thasos*, Unpublished Master Dissertation, University of Lille 3, 2016.

G.D. PAPADIMITRIOU, "Ore washeries and water cisterns in the mines of Laurion-Attica", in: K. WELLBROCK (ed.) *Cura Aquarum in Greece,* Schriften der DWhG, Band 27-2, Seigburg, 2017, 395-419.

C. PERISSORATIS, D. MITROPOULOS, " Late Quaternary Evolution of the northern Aegean Shelf", *Quaternary Research* 32 (1989), 36-50.

J. SPEIDEL, *Beiträge zur Kenntnis der Geologie und Lagerstätten der Insel Thasos*, 1929

M. VAVELIDIS, G. GIALOGLOU, E. PERNICKA, G.A. WAGNER, "Die Buntmetall- und Eisen-Mangan-Lagerstätten von Thasos", in: G.A. WAGNER and G. WEISGERBER (eds), *Antike Edel- und Buntmetallgewinnung auf Thasos, Der Anschnitt*, Beiheft 6, Bochum: Bergbau-Museum, 1988, 40-58.

G.A. WAGNER and G. WEISGERBER, *Antike Edel- und Buntmetallgewinnung auf Thasos, Der Anschnitt,* Beiheft 6, Bochum: Bergbau-Museum, 1988.

7. URBAN METALLURGY AT PISTYROS: INVESTIGATING THE CONTEXTS OF PRODUCTION

Nerantzis Nerantzis, Sabine Nodin and Stratis Papadopoulos

Introduction

The extraction of metals in northern Greece features prominently in the literature of Greco-Roman antiquity and has attracted persistent scholarly attention in the context of Classical studies. According to ancient literary sources, mining for precious and base metals focused mainly on the island of Thasos, the region of the Lekani near Krinides where the city of Philippi was established by Philip II, Mount Pangaion, the Strymon region and further to the northwest in Paionia and Mount Dysoron.[1] Interdisciplinary research combining archaeology and geological prospection was initiated in the 1970s and over time several mining and metallurgical sites have been located across this extended region.[2] Concerning the littoral area of Kavala prefecture, opposite the island of Thasos, most of these early survey findings featured in Koukouli-Chrysanthaki's seminal article "Metals of the Thasian Peraia (in Greek)" that was published in 1990 and gathered all data available during this time.[3] In recent years it became clear that attempting to understand the technical intricacies of transforming minerals into metals, should also involve an examination of the social organization and cultural context within which the communities of a certain region interacted with their natural environment as active agents within a productive landscape.

At the time of Greek colonisation of the North Aegean from around the mid-7th century BC the region between the Strymon and Nestos rivers, was inhabited by Thracian tribes described in Greek literary sources as independent *ethne* with varying degrees of political organization.[4] Colonial expeditions of the Archaic period brought Greek settlers from the Aegean islands and Asia Minor to the Thracian shores (Fig. 1) that were renowned for their riches from earlier contacts in the Geometric period.[5] The minting of silver coinage in the late 6th century BC by several cities across the region such as Thasos, Neapolis, Abdera, Maroneia and Argilos is clear proof of either direct or negotiated access to precious metals.[6] The extensive network of Thasian colonies that had direct access to mineral deposits was catalytic for increased precious metals extraction in Aegean Thrace between the 6th and 3rd centuries BC.

Although mining activity dated to the Hellenistic and Roman period has become a focus of interdisciplinary investigations on Thasos,[7] the region of Mount Pangaeon[8] and Mount Lekani[9] the actual metallurgical processes to extract silver and gold have not received adequate attention. Since mining is an activity that was continuous, often spanning many centuries, the exact dating of various extraction phases is often problematic mainly due to the obliteration of earlier evidence by more recent interventions.[10] In a few cases the shape of the galleries hints to specific extraction techniques attributed to certain periods, while pottery or clay lamps provide more dating clues. Nevertheless, in many localities the re-opening of ancient mines during the Byzantine and Ottoman period has significantly obscured issues of chronology. In similar terms the metallurgical waste that forms large heaps of slag accumulated over centuries is notoriously difficult to date as the latest phases covered or disturbed archaeological material of earlier periods. Until recently no well-dated contexts of metallurgical activity have been investigated in this region and the current case study reviews an initial attempt at addressing such technological issues from

1 Hdt. 6, 46 – 47; Thuc. 1, 100; 4, 105; Strab. 14, 5, 28; 7, 331 – 334.

2 Wagner and Weisgerber 1988; Photos *et al.* 1989.

3 Koukouli-Chrysanthaki 1990.

4 Tsiafaki 2020.

5 Tiverios 2008.

6 Picard 2006.

7 Wagner and Weisgerber 1988.

8 Vaxevanopoulos *et al.* 2017.

9 Vavelidis *et al.* 1996; 1997a; 1997b; Nerantzis 2015.

10 Stöllner 2014.

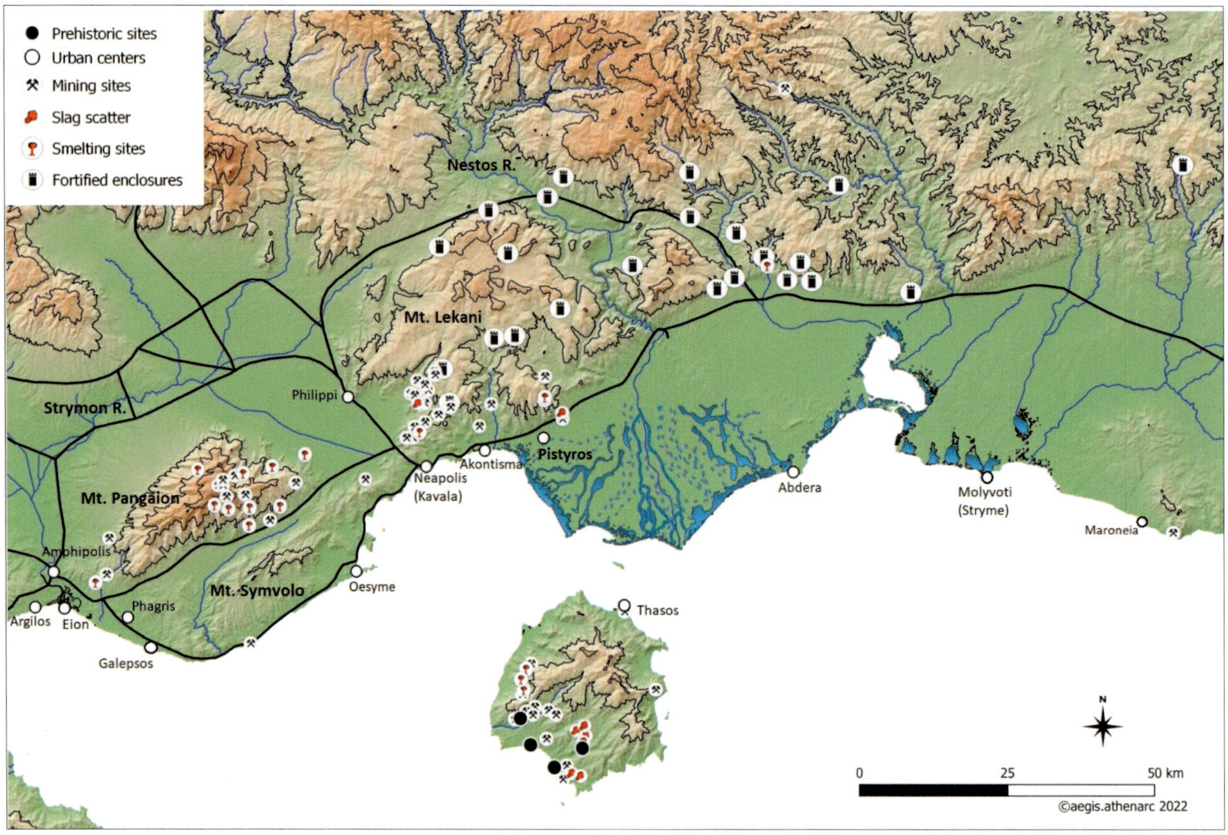

Fig. 1. Map of the north Aegean region showing major Greek colonies (designed by V. Evangelidis).

secure contexts for the first time. One of the major aims of Project *ME.Tech.NAS.* has been to investigate how metallurgical activity was associated to a complex web of relations between Greeks of the colonies, indigenous Thracian populations and Macedonians after their expansion to this mineral rich region. Information from previous archaeological projects were considered but notably, ongoing excavation (2014-2020) of a fortified Thasian settlement opposite Thasos, identified with Pistyros, provided evidence of large-scale metallurgical activities within an urban context.[11] As the evidence derives from layers dating between the late 4th and 2nd centuries BC it offers an unprecedented opportunity to study the production of metals in well-dated contexts which is a rare occurrence in the broader region.

The site and its surroundings

The ancient settlement close to the village of Pontolivado, Kavala Prefecture was discovered in 1971. Based on the findings of a rescue excavation during road construction the first excavator, Ch. Koukouli-Chrysanthaki identified the site with Pistyros.[12] She found supporting literary evidence in Herodotus' description of the march of Xerxes' army through the Thasian territories in Aegean Thrace. In this passage the description of a salty lake close to Pistyros that was the first Thasian establishment that the Persian army encountered after crossing the Nestos is in accordance with the geographic settings and geomorphological characteristics in the vicinity of the investigated site.[13] Alternative names exist in the literature mainly Πίστειρα, Βίστιρος and Βάστειρα.[14] Stephanus Byzantius mentions Vistiros, a city in Thrace[15] that is distinguished from the emporion Pistiros, which based on an inscription of the 4th century BC should be located at Vetren in Bulgaria, close to the springs of river Evros and is associated with the ethnonym Pistiritai.

11 Nerantzis *et al.* 2022.

12 Koukouli-Chrysanthaki 1972, 529; 1973a, 230-240; 1973b, 451; 1980.

13 Hdt. Z, 109-110.

14 Fragmenta Jacoby 2a, 72, frg. 10.

15 St. Byz. *Ethnika* 171 and 525.

2

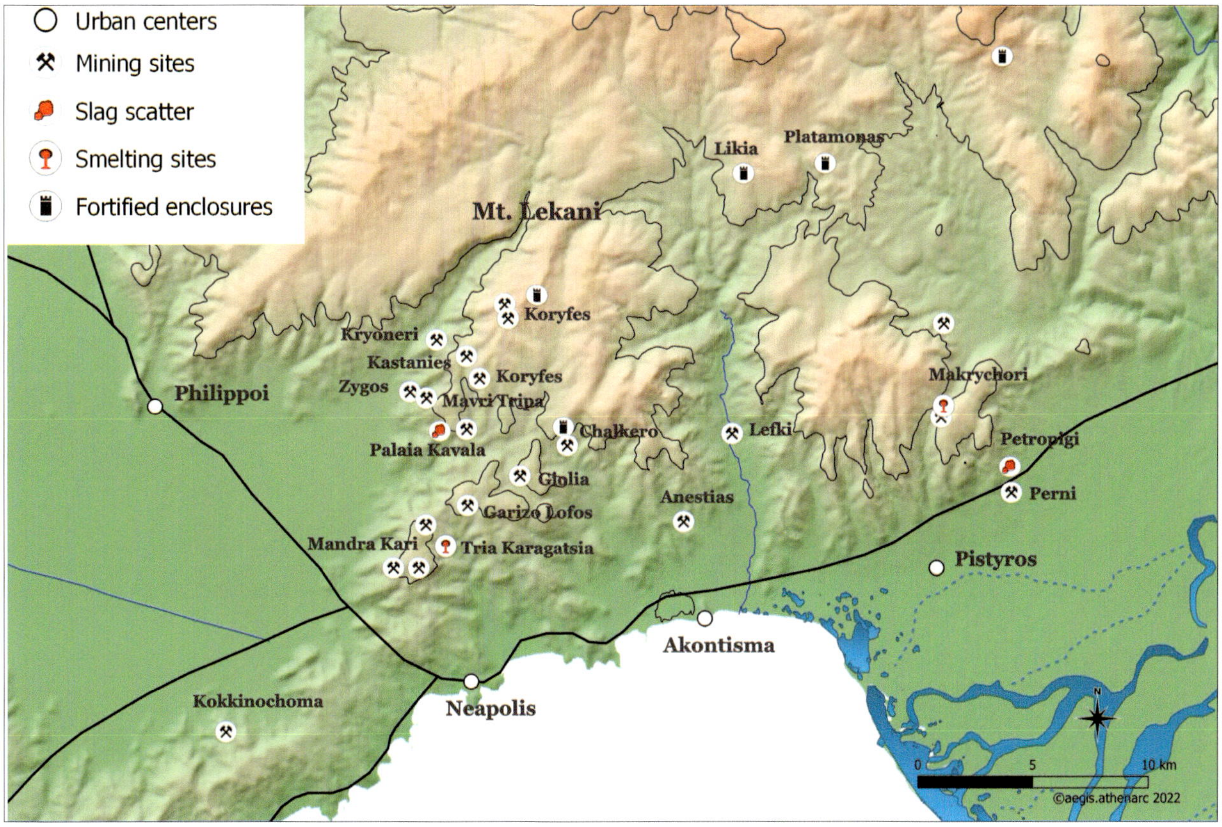

3

Fig. 2. Map of the Lekani mountain range, showing sites of mining, metallurgical activity, major urban centres and fortified enclosure sites (designed by V. Evangelidis).

Fig. 3. Mine entrance close to the village of Perni, Lekani mountain range (photo S. Nodin).

It is important to note that the site under study lies at short distance to an extensive mining zone, within a radius of 8 km, extending towards the northwest and northeast on the Lekani mountain range (Fig. 2). About 2.5 km to the NE, between the modern villages of Perni and Petropigi, there is substantial mining evidence dated tentatively to the Classical and Hellenistic periods. An underground mine with a single entrance and narrow galleries of square profile is located about 1 km west of the modern village of Perni (Fig. 3). A similar underground mining shaft with a stepped entrance is located about 500 m NE of Petropigi.[16] The tool-marks on the galleries' walls which were made by miner's picks and hammers as well as the characteristic profiles suggest a relative dating for these workings to Classical antiquity. Scatters of surface slag have been noted close to Petropigi and several samples were taken for analysis as comparative material with that excavated at Pistyros.

The valleys of Lefki and Chalkero allowing access to the central and western sides of the Lekani massif have yielded further evidence for mining and several concentrations of slag. Near the village of Ano Lefki, numerous mining galleries of square profiles and surface slag deposits are reported in previous studies.[17] At a locality called Mylos Deli, northwest of the village of Ano Lefki, a tower has been erected sometime around the end of the 6th century BC, possibly for the coordination of mining activity in the area.[18] As suggested by Koukouli-Chrysanthaki based on its architectural features and the relative dating, the tower could be related to the settlements

16 Nerantzis *et al.* 2022.

17 Koukouli-Chrysanthaki 1990; Photos *et al.* 1989.

18 Zannis 2014, 191.

of the Thasians, near Nea Karvali, i.e., Akontisma and possibly also Pistyros.[19]

Previous geological research has revealed that the ore mineralisation of the Lekani mountain range occurs as carbonate replacement bodies and veins in marbles and schists.[20] Thus three main types of mineralisation have been identified, namely Au bearing Fe-Mn oxidised bodies, Pb-Ag and/or Zn bearing Fe-Mn oxidized bodies and Au bearing pyrite-arsenopyrite and pyrite-chalcopyrite bodies.[21] Gold is present in native form in the iron oxides and in some cases within the arsenopyrite and chalcopyrite. Silver is present in native form and in sulphidic compounds, mainly in the rich in galena and jarosite ore.[22]

Based on the findings from the systematic excavation initiated in 2014 it has been confirmed that Pistyros was established in the end of the 7th-beginning of the 6th century BC and was abandoned sometime during the 1st century BC.[23] The finding of local, handmade pottery might suggest that the site was founded following mutual agreements between the Thasians and the local Thracian tribe of the Sapaioi. Parts of the street network and four residential quarters of the central and southeastern sectors have been excavated between 2014 and 2019.[24] The entire perimeter of the fortification wall has been investigated, featuring six square towers and two gates, a major one at the SE and a secondary one, of smaller dimensions, at the northern corner. Remains of an impressive stone-paved street came to light commencing at the gate and following an inclined course for some meters, leading out of the city towards the southeast. The fortification wall displays two major building phases, the first of which, showing remarkable similarities to the wall of Thasos.[25] The eastern, northern and western sections display characteristics of the first building Phase of ashlar masonry using the isodomic technique with massive, rectangular marble slabs. Several slabs bear inscribed names with the most commonly occurring those of Ἐπιγ(ένης) and *Γλαῦ(κος)* the latter name being common in Thasos.[26]

A hoard of 55 Archaic silver coins of Thasos and Neapolis was found in 1971 during the first investigation under a slab of the external eastern side of the wall at the NE corner.[27] The Thasian coins of the earliest type of Silenus abducting nymph are 10 staters and 29 smaller denominations and date between 520 and 510 BC. The Neapolis coins with gorgon's head are 12 staters and 4 smaller denominations dating between 500 to 480 BC.[28]

Contexts with metallurgical finds

During the course of ongoing excavation at Pistyros, particularly at the residential quarters of the central and southeastern sectors, a large volume of metallurgical residues has been retrieved from most of the investigated contexts. Recording of these finds started in 2016 and up to 2020 about 1,220 kg of slag and related residues have been registered. This is unusual for an urban context since such activities were more often taking place closer to the mines or in areas close to the walls of a city either internally or externally. For instance, excavations at the Archaic-Classical city of Apollonia Pontica on the Black Sea coast revealed a complex of metallurgical furnaces at the northern end of the isthmus immediately outside the city.[29] They are dated to the second half of the 6th century BCE and testify to the existence of an *extra muros* industrial quarter dedicated to bronze metallurgy, where some 30 furnaces were found.[30] Substantial evidence for metallurgy, mostly in form of slag deposits are reported from various spots in the city of Thasos[31] and to a lesser extent from a metalworking workshop at Argilos.[32] For the case of Pistyros an interpretation of the findings related to metallurgy is attempted both in terms of their spatial distribution and stratigraphic associations as well as by a targeted instrumental analysis to account for the actual metallurgical processes prevalent on site. Looking at the distribution of such finds we noted that squares ΒΨ, ΓΣ and ΕΗ yielded the largest volumes (Fig. 4). Such accumulations at a central area of the site could suggest either a direct deposition of waste near the workshops, or it could be proof of a secondary deposition pattern that is still under investigation. For the moment no space has been found that could be identified with certainty as a workshop. Yet a few stone blocks bearing signs of high

19 Koukouli-Chrysanthaki 1980.

20 Vavelidis *et al.* 1996; Fornadel *et al.* 2011.

21 Vavelidis *et al.* 1997a, 97.

22 Vavelidis *et al.* 1996, 553.

23 Papadopoulos and Zannis 2022

24 Papadopoulos and Zannis 2022; Papadopoulos*et al.* in press.

25 Grandjean *et al.* 2011.

26 Papadopoulos and Zannis 2022, 556.

27 Koukouli-Chrysanthaki 1973a, 235.

28 Oikonomidou 1990.

29 Damyanov and Panayotova 2020, 259.

30 Baralis *et al.* 2016, 159.

31 Pichot 2010.

32 Perreault and Bonias 2012.

4

vitrification and a thin layer of slag on one surface and pieces of vitrified clay lining should belong to furnace debris and therefore we presume that the workshops were probably somewhere in the vicinity of squares ΒΨ and ΓΣ.

A preliminary, macroscopic examination of the residues provided some first clues concerning the main metallurgical processes represented on the site. Thus, the great majority of finds consists of smelting slag with typical ropey flow textures deriving from the reduction of various mineral ores. In the areas where a stratigraphic sequence could be followed, slag deposition appears associated to a single occupational phase, dating from the last quarter of the 4th and covering the 3rd and early 2nd century BC. In the southeastern sector of the sanctuary (squares ΓΗ and ΓΩ), buildings of this Phase have been revealed. The upper layer that covered the architectural remains at square ΓΗ contained 7.4 kg of slag while the layer directly below has yielded 2.4 kg of slag scattered above a floor surface of compacted red clay that has been revealed across an area of 2 x 1 m. In the neighbouring square, ΓΩ where excavation continued at deeper levels a courtyard with two marble pedestals have been revealed. At this area numerous pieces of slag weighing up to 8.2 kg have been recovered.Several other contexts yielded relatively larger volumes of slag as is the case with the large building excavated in squares ΔΛ, ΔΜ and ΕΗ (Fig. 4). From square ΔΜ a total of 44 kg were counted from across the entire excavated surface and 25 kg from the western part of the building investigated in square ΔΛ. Larger volumes were retrieved from square ΕΗ reaching 102 kg, with a higher concentration at the northern part of trench

5

Fig. 4. Spatial distribution of slag across the site.

Fig. 5. The circular stone-built feature, K1 with clay backing externally and surrounding walls (trench EH3).

EH3. These were found around a circular stone-built feature, labelled K1, with clay backing externally that was also filled with slag (Fig. 5). There is a wall of earlier date that stretches under the northern side of K1 while at the south there is a parallel wall (T1) seemingly contemporary to this circular feature. The accumulation of stones to the SW of K1 appears to be part of the collapsed southern wall T1. The area west of K1 towards wall T5 contained 10.5 kg of slag, while the area immediately to the south contained 3.7 kg. The layer overlying the circular feature was filled with stone debris and a quantity of slag weighing 13.8 kg. From its interior a total of 14.4 kg of slag were recovered along with some pieces of charcoal and masses of red clay. This concentration of waste might suggest that the area was a dumping space for metallurgical residues. Notably the pieces of slag

6

7

8

Fig. 6. Square BΨ: The street surface was partly covered by the slag deposit that extends below wall T2 of the eastern cluster.

Fig. 7. The extensive slag deposit unexcavated below wall T2. This layer of compacted slag covered the entire area of the stone-paved floor with drainage channel (foreground) in the eastern occupational cluster (trench BΨ3).

Fig. 8. Slag cakes and furnace debris in situ at the edge of the stone-paved area of the eastern cluster (trench BΨ3).

recovered are mostly broken, of small sizes not more than 5 cm, probably due to deliberate fragmentation for a possible post-smelting treatment. There was no evidence of high heating of the soil around or within K1. Although the actual function of this feature has not been verified yet it is notable that most of the smallest pieces were retrieved from its interior.

Square BΨ yielded the largest volumes of slag recorded so far on site, in the order of 536 kg. The main features of this area include a street of NW-SE orientation, dividing two building clusters, dating from the late 4th to the 3rd c. BCE (Fig. 6). The dark brown deposit covering part of the street contained a large amount of slag as well as charcoal spread over a layer of varying thickness between 20 and 50 cm. Trench BΨ3 contained the largest volume of slag from a single trench (447 kg) in a compact layer below wall T2, across the space to the east, as well as within the deposit covering the street. A trial trench opened east of T2 has revealed that this compact layer of slag is about 0.20-0.50m thick and predates the foundation of this wall that was erected at a later stage (elevation 9.00-8.90m). Upon removal of this layer a stone paved floor was revealed (elevation 8.70 m) featuring a stone-built drainage channel for directing water towards the street (Fig. 7). Besides slag, some vitrified and slagged stones, clay lining and furnace wall fragments have been recovered from the same layer covering this stone paved area (Fig. 8).

Preliminary dating of the slag deposit of square BΨ was attained by numismatic evidence. In a context on the northern part of the open area delimited by walls T1 and T2 a bronze coin of Abdera was recovered, struck between 311 and 281 BC.[33] Since this context overlies the slag deposit, the coin offers a *terminus ante quem* for slag deposition. Moreover, in a context within the slag deposit, below wall T2 a bronze coin of Alexander III dated 336-323 BC provides further dating clues.[34] Its presence here could date the metallurgical activity, or indeed the dumping of slag,

33 Chryssanthaki-Nagle 2007, 209-222.

34 Gatzolis 2012, 382-385.

9

11

10

Fig. 9. Lead oxide cake (litharge) recovered at square ΓΣ (PIST38-ΓΣ 1 #2 AE11).

Fig. 10. Spatial distribution of speiss and litharge across the site.

Fig. 11. Piece of speiss (Fe/As/Sb) from square ΓX (PIST32-ΓX 2 #2 AE8).

to the late Classical/early Hellenistic period. Such dating is also corroborated by a preliminary study of the pottery that includes some typical examples of west slope ware.

Besides slag, some heavy, pinkish-grey metallurgical residues were noted in several contexts across the site (Fig. 9). These are solid blocks of lead-oxide referred to as "litharge", usually interpreted as derivatives from a cupellation process for the separation of lead from silver. This was achieved in a secondary step following the smelting of a silver bearing lead ore. In total 23 pieces of "litharge" have been retrieved displaying two apparent concentrations of 6 to 7 pieces, at squares BΨ and ΓΩ and isolated finds in four other squares (Fig. 10). Another characteristic type of residue was noted during the recording phase, which appears rusty on its exterior surface and upon breaking reveals a silvery lustre (Fig. 11). This is commonly referred to as speiss in modern metallurgical terminology.[35] The finding of these rich in Fe/As/Sb residues testifies to the possible extraction of precious metals from polymetallic or low-grade ores such as jarosite or arsenopyrites as alternatives to the dry silver ores such as argentite and cerargyrite or the rich in silver galena. Any interpretation though is still premature before

35 Bachmann 1982, 29; Thornton *et al.* 2009.

Fig. 12. Buildings of the sanctuary area in the southeastern sector. In square ΓΩ the well (w) and the large marble pedestal (p) are indicated.

completion of the laboratory analyses, which are currently ongoing. These finds were more dispersed across space in the southern and central squares and add up to 9 pieces.

The sanctuary area (squares ΓΨ, ΓΩ, ΓΖ, ΓΗ) where a concentration of litharge and speiss was noted, is characterised by a courtyard of the 3rd century BC featuring two carefully carved marble pedestals (Fig. 12). At least six buildings, including a temple, have been excavated in this area belonging to two major building phases. East of the pedestal a wide wall extends with a slight inclination towards the NE and appears to be enclosing a large area, which continues into the neighbouring square to the north. In the southeastern corner of square ΓΩ, a stone built well has been revealed containing a large amount of pottery and some metallurgical waste. Notably among these finds there was a piece of speiss of large proportions measuring 24 x 18 cm. It is notable that on one side it displays a pronounced curvature that would form when at some point during the process liquid speiss was in contact with a curved surface. Within the overlying context a bronze coin of Antigonus II Gonatas struck between 277 and 239 BC[36] offers some dating evidence for this important metallurgical find that could have been deposited sometime before the mid-3rd century BC.

The litharge and speiss from this area come from several contexts with the majority deriving from the courtyard of the sanctuary and less so among the 3rd century buildings. These finds testify to primary and secondary precious metals extraction processes possibly taking place in the vicinity. Although they were found at varying depths between 8.30 to 7m ASL they all belong to the same chronological phase. In terms of their size, litharge pieces are about 6 to 7 cm in average with two larger examples reaching 15 cm. The speiss fragments are about 10 cm in average with the largest example from the well deposit reaching up to 24 cm in length.

Another concentration of 6 litharge pieces comes from the slag deposit in square ΒΨ. These are mostly fragmented with one exception of a more complete

36 Kremydi-Sicillianou, pl. XLVII, 1010-1015.

13

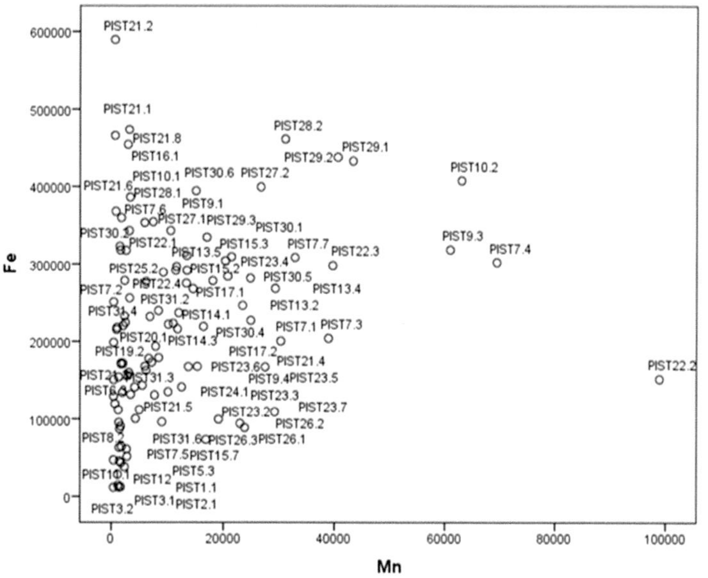

Fig. 13. PCA plot based on pXRF data. Slag, speiss and litharge from Pistyros taking as variables Fe and Mn.

14

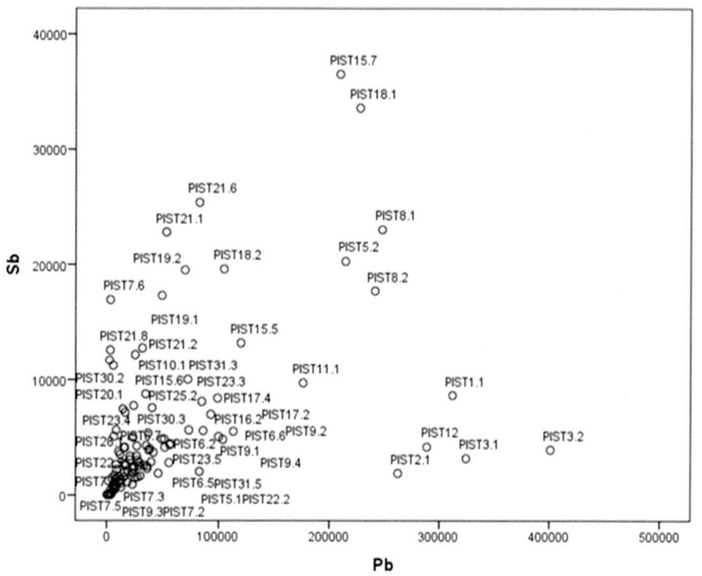

Figure 14. PCA plot based on pXRF data. Slag, speiss and litharge from Pistyros taking as variable Sb and Pb.

cake displaying a roughly circular perimeter with ridges on its upper surface caused during cooling. Two similar cake fragments come from squares ΑΣ and ΓΣ that also display the characteristic rippled effect on the upper surface.

Preliminary results of pXRF analysis

In order to better characterize the metallurgical processes, a series of instrumental analyses is currently under way on selected finds from contexts belonging to the best-documented phase excavated so far, that of the Hellenistic period. Before attempting any characterisation of the chaine operatoire for metals extraction at Pistyros it is important to focus on one particular feature of the metallurgical finds, and that is their size. Since the average slag size could account for any intentional treatment after smelting, this information could be indicative of a specific step in the metal extraction process. For the case of Pistyros the average slag size ranges between 1 and 5 cm. In absolute figures out of almost 12,300 counted pieces about 228 are larger than 10cm, representing only 1.8% of the total assemblage. This means that small pieces, often broken, represent the great majority while larger ones, although present, are more rare. Looking at the distribution of the larger pieces surpassing 10 cm in size we have noticed the following pattern. In 7 squares there are very low concentrations of large pieces below 1%. In 9 squares such pieces represent 1-5%. Finally, in square ΑΔ they represent 6% and in ΓΠ up to 11%. Based on the above, the small average size is most likely the result of breaking up either on site or elsewhere and this most probably shows intent for further treatment. The above considerations dictated the sampling methodology in order to select representative samples during the stages of in-situ and laboratory analyses.

The first stage of the analytical methodology involved non-invasive X-ray fluorescence spectroscopy with a Niton portable XRF performed on site.[37] Over one hundred measurements were taken, and the results suggest that all three types of residues namely slag, litharge and speiss fall within one major group. The PCA distribution taking as variables Fe and Mn shows one rather homogeneous group with a few outliers (Fig. 13). The correlation of these elements could be attributed to the nature of the Fe/Mn ores that were smelted. When Sb and Pb are taken as variables most samples plot in a compact cluster at the low

37 The ongoing instrumental analyses are performed in collaboration with E. Photos-Jones and results will be reported elsewhere.

concentration region while more outliers are noted (Fig. 14). These outliers derive from the circular, stone-built feature (K1) of trench EH3, which needs further investigation. Future work will focus on discerning the Sb and Pb correlation and how their relative ratios could be reflecting the extracted metal(s) on site.

With the data available so far, we could suggest that the ore smelted at Pistyros, containing Fe/Mn/Pb/Sb, is likely to have derived from the mines at Perni and Petropigi. We suggest that in Pistyros there is a continuity in working the same ores whether in the Classical – for the Thasians – or in the Hellenistic – for the Macedonian kings. If that is the case the city must have had an acknowledged identity as a metal producing area within its region and possibly also beyond.

Discussion and prospects

Attempting to understand the organisation of mining and metal production of the Thasian state, comprising the island and its mainland colonies we should take into account issues of access to raw materials, infrastructure and workforce availability from the 6th to the 2nd century BC. Considering the situation across the whole region colonized by the Thasians, it seems that only the indigenous settlements situated on the littoral were replaced by Thasian foundations such as Neapolis or Oesyme.[38] It is also indicative that Thracian settlements such as the one at Eleutheroupolis by the foothills of Mount Pangaeon, Hill 133, and the site near Gazoros by the Strymon, in the vicinity of Thasian colonial foundations, remained inhabited and unaltered.[39] It is through this coexistence that exploitation of natural resources should be perceived, and the archaeological record provides some supportive numismatic evidence.

The earliest coinages produced in littoral Thrace date to the late 6th century BC, and they were minted by the Greek cities as well as the local Thracian tribes such as the Bisaltes, Hedonoi, Orescii and others.[40] The vast majority of issues are stater size or larger, although smaller denominations do exist.[41] The cities of Berge, Thasos, Neapolis, Galepsos, and Ennea Hodoi that were situated at crucial trading crossroads issued staters of ca. 10 g and fractions on the duodecimal system.[42] A number of scholars have suggested that staters minted by cities and local tribes, and triple-staters minted by local tribes, could be used as bullion, a commodity for export[43] or to pay tribute to the Persians and later to the Athenians.[44] Based on a recent study of more than 200 coins of the centaur and nymph "Thraco-Macedonian" series it has been suggested that these were struck in the area east of Mount Pangaion, as part of tribute paid to the Persian King.[45] In addition, that the variety of depictions could be interpreted as actually indicating the provenance of the metal used, while the legends, where they exist, concern the tribe that extracted the metal.[46]

Interpretations for the simultaneous appearance of these early coinages range from monetary and military alliances to a less formal co-operation in the use of mines or mints. Cooperative mining upon arrangement between parties could account for the concurrent minting of tribal, civic and royal coinages across this region.[47] Such a situation could explain how the small numbers of colonists succeeded in appropriating and exploiting the mineral resources laying in Thracian territory, presumably in cooperation with numerous indigenous miners and labourers in fuel preparation, ore beneficiation and enrichment as well as smelting at several locations.

Since the metallurgical residues from Pistyros derive from the extraction of complex polymetallic ores dating to the Late Classical/Hellenistic periods it is important to highlight some major historical events of this timeframe in association to the wealth generated by the exploitation of such resources. During the reign of Philip II the eastward expansion of the Macedonian kingdom brought the subjugation of the territory between the Strymon and Nestos under Macedonian authority.[48] The objective of founding Philippi in 356 BC at the site of the Thasian settlement of Krinides was possibly to take control of the neighbouring gold and silver mines and to establish a garrison at a strategic passage. The site controlled a major route between Amphipolis and Neapolis, part of the great royal highway that crossed Macedonia from east to west and which was reconstructed later by the Roman Empire as the Via Egnatia. Diodorus Siculus mentions that the mines close to Krinides were not producing a large output and by developing better infrastructure Philip II managed to increase the revenue from gold

38 Koukouli-Chrysanthaki 1980.

39 Baralis 2009, 116.

40 Picard 2006.

41 Picard 2000; Mitchiner 2014; Psoma 2015.

42 Psoma 2015, 175.

43 Kroll 2011.

44 Picard 2000.

45 Tzamalis 2012.

46 Tzamalis 2011, 75.

47 Kosmidou 2011, 443.

48 Koukouli-Chrysanthaki and Bakirtzis 1997; Sève 2000.

mining substantially.[49] Pistyros, that was strategically located close to the eastern Lekani mines and the commercial and military routes witnesses renewed economic prosperity under Macedonian rule. An inscription of the Hellenistic period found on site and is currently under study provides proof to the existence of a cult of king Philip II.

At the present level of investigation Pistyros presents us with unresolved questions: There is nearly 1.2 tons of metallurgical waste, across the site and at depth. There are so far no excavated furnace remains, although sporadic furnace wall fragments have been found. There are so far no artefacts (jewellery or coinage) that would allow us to say with confidence that the inhabitants of Pistyros are producing bullion for one of the Macedonian towns which minted coinage. Yet they come across as confident metal manufacturers: first, they appear to have had ready access to the eastern Lekani mountains with known precious metals deposits and secondly they appear to have been smelting that ore and processing it further, within their town's boundaries, over a long period of time (before the arrival of the Macedonians) as can be deduced from amounts of slag tentatively dated to the Classical period. This suggests that they must have enjoyed a period of stability under the protection of a ruling authority which valued their expertise and benefited from it.

Due to their abundance and well-dated contexts the metallurgical finds from Pistyros could act as reference material for a comparative study of metallurgy in the broader region that is attested on upland, mostly undated contexts. The ongoing analysis of Pistyros metallurgical residues is expected to elucidate further the model for metals extraction on site. This could in turn be used to evaluate its inhabitants' knowledge in metallurgical technology and the site's importance in this productive sector that was fundamental for its economic prosperity between the 4th and 2nd centuries BC.

Abbreviations

Hdt. 6, 46 – 47 = Herodotus, *Histories*
Thuc. 1, 100; 4, 105 = Thucydides, *History of the Peloponnesian War*
Strab. 14, 5, 28; 7, 331 – 334 = Strabo, *Geographica*
Diod. Sic. XVI 8.6-7 = Diodorus Siculus, *Library of History*

49 Diod. Sic. XVI 8.6-7.

Bibliography

H.-G. Bachmann, "The Identification of Slags from Archaeological Sites", Institute of Archaeology, London, 1982.

A. Baralis, "The Chora Formation of the Greek Cities of Aegean Thrace. Towards a Chronological Approach to the Colonization Process", in: P. Guldager Bilde, J. H. Petersen (eds), *Meetings of Cultures in the Black Sea Region: Between Conflict and Coexistence*. Aarhus, Aarhus University Press (Black Sea Studies 8), 2009, 101-130.

K. Chrysanthaki-Nagle, "L'Histoire Monetaire D'Abdere En Thrace (VIe s. av. J.-C.- IIe s. ap. J.-C.)", *Meletemata* 71, Athènes, Centre de Resèche de l'Antiquité Grecque et Romaine, Fondation National Hellénique de la Recherche Scientifique. 2007.

P.T. Craddock, *Early Metal Mining and Production*, Smithsonian Institution, Washington D.C., 1995.

M. Damyanov and K. Panayotova, "Apollonia Pontica and the South-Western Black Sea Coast in the Thracian and Antique Periods (Late 2nd Millennium BC to Early 1st Millennium AD)", in: R. Kraus, E. Pernicka, R. Kunze, K. Dimitrov and P. Leshtakov (eds), *Prehistoric Mining and Metallurgy at the Southeast Bulgarian Black Sea Coast*, Ressourcenkulturen 12, Tubingen University Press, 2020, 247-294.

A.P. Fornadel, P.G. Spry, V. Melfos, M. Vavelidis and P. Voudouris, "Is the Palea Kavala Bi-Te-Pb-Sb ± Au district, northeastern Greece, an intrusion-related system?", *Ore Geology Reviews*, 39 (2011), 119-133.

Ch. Gatzolis 2012 "Επανεξέταση της χαλκής νομισματοκοπίας του Αλεξάνδρου Γ' και του Κασσάνδρου υπό το πρίσμα νέων ανασκαφικών και θησαυρικών δεδομένων", in: M. Tiverios, P. Nigdelis and P. Adam-Veleni (eds), Θρεπτήρια: Μελέτες για την Αρχαία Μακεδονία, Θεσσαλονίκη, 2012, 380-387.

G. Gounaris and E. Gounaris, *Philippi: Archaeological Guide*, Thessaloniki, 2004.

Y. Granjean, M. Wurch-Kozelj and T. Kozelj, "Le rempart de Thasos", Études *Thasiennes XXII*, Ecole Francaise d'Athenes, 2011.

E. Kosmidou, "Macedonian and Thracian Relations in Northern Greece after the Persian Wars with Particular Reference to the Coinage and Politics of Alexander I" in: *Travaux de Symposium International les Livre. La Roumanie. L'Europe*, Tome IV. Bucarest, Editeur Bibliothèque de Bucarest, 2011, 439-452.

Ch. Koukouli-Chrysanthaki, "Ανατολική Μακεδονία, Νομός Καβάλας, Πίστυρος", Αρχαιολογικόν Δελτίον 27, Χρονικά Β'2 (1972), 529.

Ch. Koukouli-Chrysanthaki, "Εἰδήσεις ἐκ τῆς Θασίων Ἠπείρου", Αρχαιολογικά Ανάλεκτα εξ Αθηνών 6 (1973a), 230-240.

Ch. Koukouli-Chrysanthaki, "Ανατολική Μακεδονία, Νομός Καβάλας, Πίστυρος", Αρχαιολογικόν Δελτίον 28, Χρονικά Β' 2 (1973b), 451.

Ch. Koukouli-Chrysanthaki, "Οι αποικίες της Θάσου στο Βόρειο Αιγαίο, Νεώτερα ευρήματα", in: Πρακτικά Α' Συνεδρίου "Η Καβάλα και η περιοχή της", Καβάλα, 1980, 309-325.

Ch. Koukouli-Chrysanthaki, "Τα 'μέταλλα' της Θασιακής Περαίας", in: *Μνήμη Δ. Λαζαρίδη: Πόλις και Χώρα στην Αρχαία Μακεδονία και Θράκη, Πρακτικά Αρχαιολογικού Συνεδρίου (Καβάλα 1986)*, Ελληνογαλλικές Έρευνες Ι. Θεσσαλονίκη, 1990, 493-514.

Ch. Koukouli-Chrysanthaki and Ch. Bakirtzis, *Philippi*, Ministry of Culture, Archaeological Receipts Fund, Athens, 1997.

S. Kremydi-Sicillianou, *Sylloge Nummorum Graecorum, Greece II, The Alpha Bank Collection, Macedonia I, Alexander I – Perseus*, Athens, 2000.

J.H. Kroll, "Minting for Export: Athens, Aegina and Others", in: Th. Faucher, M. Chr. Marcellesi and O. Picard (eds) *Nomisma. La circulation monetaire dans le monde grec antique. Actes du colloque international*, Athenes, 14–17 avril 2010, BCH suppl. 53. Athenes: Ecole francaise d'Athenes, 2011, 27-38.

M. Mitchiner, *Ancient trade and early coinage*, Hawkins Publications, London, 2004.

N. Nerantzis, *Rhesos' Gold, Heracles' Iron: the archaeology of metals exploration in Northeast Greece*, Potingair Press, Glasgow, 2015

N. Nerantzis, S. Nodin and S. Papadopoulos, "The organization of mining and metal production in Aegean Thrace from the Archaic to the Roman period", in: F. Hulek and S. Nomicos (eds) *Ancient Mining Landscapes, Panel 4.2, Archaeology and Economy in the Ancient World*, Vol. 25, Heidelberg, Propylaeum, 2022, 65-84.

M. Οικονομιδου, 1990 Αρχαϊκός Θησαυρός Αργυρών Νομισμάτων από το Ποντολίβαδο (1971), *Μνήμη Δ. Λαζαρίδη: Πόλις καί Χώρα στην Αρχαία Μακεδονία και Θράκη, Πρακτικά Αρχαιολογικού Συνεδρίου (Καβάλα 1986)*, Ελληνογαλλικές Έρευνες Ι. Θεσσαλονίκη, 533-540.

S. Papadopoulos, S. Dittopoulou, Ch. OIkonomidis and E. Fergadiotou, "Συστηματική ανασκαφή Πίστύρου: η πρώτη πενταετία", *Το Αρχαιολογικό Έργο στη Μακεδονία και τη Θράκη* 33 (2019-2020), in press.

S. Papadopoulos and A. Zannis, "Η Πίστυρος της αιγαιακής Θράκης", *Το Αρχαιολογικό Έργο στη Μακεδονία και τη Θράκη* 30 (2016), 2022, 553-564.

J. Perreault and Z. Bonias, "After Zagora-Andrian colonization in the northern Aegean: the case of Argilos", *Mediterranean Archaeology* 25, (2012), 259-271.

E. Photos, Ch. Koukouli-Chrysanthaki, R. F. Tylecote and G. Gialoglou, "Precious Metals Extraction in Palaia Kavala, NE Greece, An Archaeometallurgical Attempt to Locate Skapte Hyle", in: *International Old World Archaeometallurgy Conference, Der Anschnitt*, Heidelberg: Max Plank Institut fur Kern Physik, 1989, 179-190.

O. Picard, "Monnayages en Thrace a l'epoque achemenide", in: O. Cassabone (ed.), *Mecanismes et innovations monetaires dans l'Anatolie achemenide, Numismatique et Histoire*, Actes de la Table Ronde Internationale d'Istanbul, 22–23 mai 1997, Varia Anatolica XII. Istanbul: Institut francais d'etudes anatoliennes, 2000, 239-252.

O. Picard, "Mines, monnaies et impérialisme: conflits autour du Pangée (478-413 av. J.-C)", in: A.M. Guimier-Sorbets, M.b. Hatzopoulos and Y. Morizot (eds), Rois, cités, nécropoles: institutions, rites et monuments en Macédoine, *Meletemata* 45, Paris: de Boccard, 2006, 269-283.

V. Pichot, "Rapports 2009. Thasos, prospection minière", *BCH* (2010) 134: 519-522.

S.E. Psoma, "Did the so-called Thraco-Macedonian standard exist?", in: U. Wartenberg and M. Amandry (eds), ΚΑΙΡΟΣ Contributions to Numismatics in *Honor of Basil Demetriadi*, The American Numismatic Society, New York, 2015, 167-190.

M. Sève, "De la naissance à la mort d'une ville: Philippes en Macédoine (IVe siècle av. J.-C.-VIIe siècle apr. J.-C.)", *Histoire Urbaine* 1, 2000, 187-204.

T.R. Stöllner, "Chapter 7. Methods of Mining Archaeology (Montanarchäologie)", in: W.B. Roberts and C.P. Thornton (eds), *Archaeometallurgy in Global Perspective. Methods and Syntheses*, Springer Science and Business Media New York, 2014, 133-159.

C.P. Thornton, Th. Rehren and V.C. Pigott, "The production of speiss (iron arsenide) during the Early Bronze Age in Iran", *Journal of Archaeological Science* 36, 2009, 308-316.

M. Tiverios, "Greek Colonisation of the northern Aegean", in: G. R. Tsetskhladze (ed.), *Greek Colonization, An Account of Greek Colonies and other Settlements Overseas*, Leiden-Boston, Brill, 2008, 1-154.

D. Tsiafaki, "The Northern Aegean", in: F. de Angelis (ed.), *A Companion to Greeks Across the Ancient World*, Willey-Blackwell, 2020, 409-430.

A.R.A. Tzamalis, "Monnaies "thraco-macédoniennes" : quelques observations sur les monnaies au centaure et à la nymphe", *BCH Supplement* 53 (2011), 67-77.

M. Vavelidis, G. Gialoglou, G.A. Wagne and V. Melfos, "Goldbrube in Palaea Kavala/ Griechenland: Entdeckung von Skaptehyle?", *Erzmetall* 49, No 9 (1996), 547-554.

M. Vavelidis, G. Christofidis and V. Melfos, "Αρχαία μεταλλευτική δραστηριότητα στην ευρύτερη περιοχή Καβάλας-Φιλίππων", in: Αρχαία Ελληνική Τεχνολογία, Πρακτικά του 1ου Διεθνούς Συνεδρίου, Θεσσαλονίκη, 1997a, 95-101.

M. Vavelidis, G. Christofidis and V. Melfos, "Ancient gold mines near Krinides in Philippi area (Macedonia, N. Greece)", *Archaeometrie und Tenkmalpflege*, Wien, 1997b, 84-86.

M. Vaxevanopoulos, M. Vavelidis, D. Malamidou and V. Melfos, "Ανασκαφική έρευνα σε αρχαία μεταλλεία και μεταλλουργικές θέσεις του όρους Παγγαίου (Ασημότρυπες, Βαλτούδα)", *Το Αρχαιολογικό Έργο στη Μακεδονία και τη Θράκη* 26

(2012), 2017, 547-552.

G.A. Wagner and G. Weisgerber, *Antike Edel- und Buntmetallgewinnung auf Thasos, Der Anschnitt*, Beiheft 6, Bochum: Bergbau-Museum, 1988.

A.G. Zannis, Le pays entre le Strymon et le Nestos: géographie et histoire (VIIe-IVe siècle avant J.-C.), *Meletemata* 71, Athènes, 2014.

8. SLAGS REINVENTING THEMSELVES: FROM METALLURGICAL WASTE TO INGREDIENTS IN PHARMACOLOGICAL PREPARATIONS

Effie Photos-Jones

Abstract

Ancient Greeks had a few words to denote metallurgical slag, including *scoria*, (σκωρία) the word that came to dominate the metallurgical vocabulary for this material. Slags are integrally associated with metal-making practices (or indeed glassmaking) but, in antiquity, the scoria of some metals (lead, iron, silver) featured as ingredients in pharmacological preparations. It is not clear how metallurgical slags *moved* from the sphere of metallurgy to the sphere of pharmacology. Was the scoria of these metals "reused" metallurgical slag or was it prepared independently and in its own right?

The materials identified in antiquity as scoria of lead, iron, silver almost certainly emerged out of high temperature furnace/hearth-based processes. Compositionally they must have consisted of a combination of metal oxides, but some/all must also have contained (some) silicates if their name, implying viscosity/fluidity, was to be justified. Visually, they must have *looked* like a slag, perhaps with a ripple surface effect.

This paper sets the framework for further discussions on high temperature-derived ingredients used in pharmacological preparations, which have for long gone under-the-radar; focusing on their presence in the texts will likely create an incentive to begin looking for them in the field. I draw all references from the *Thesaurus Linguae Graecae* digital database.

Setting the scene

Metallurgical slag makes a routine entry in the material culture of antiquity from the Bronze Age onwards. Slag has been the main vehicle driving archaeo-metallurgical investigations on account of (a) its surviving amount vastly outnumbering that of ore and/or metal at any given site, and (b) its pivotal role in providing information on both metal (product) and ore (raw material). Studies of ancient slag at any site have largely focused on the elucidation of their chemical/mineralogical composition and physical properties, the aim being to elucidate the type of metallurgical activities and the key metals extracted and worked. We may know a lot about ancient slags, and from a metallurgical perspective, but we are not sure what the Greeks called the slags they made, in a specific way, that is, based on the processes that generated them (i.e., smelting, melting, alloying etc).

Contrary to the wealth of information arising from the field, references to metallurgical slag in the ancient texts are sparse and highly dispersed. Slags of four workable metals feature under the name of *"scoria"* (σκωρία):[1] *scoria siderou* (iron),[2] *scoria chalkou* (copper),[3] *scoria molybdou* (lead),[4] and *scoria argyrou (silver).*[5] But there are more names associated with each individual metal, including *tryga/tryx (τρύγα/τρύξ)* for iron;[6] *elkysma (ἕλκυσμα)* for silver;[7] *chrysanthion* (χρυσάνθιον) for copper.[8]

There is also the word *katharma (κάθαρμα)* being non-specific to any metal in particular and denoting something "*which is thrown away in cleansing, residuum of ore after smelting.*[9] Strabo[10] notes that the "*katharma*" of gold is defined as "electrum"*: "when*

1 http://stephanus.tlg.uci.edu/lsj/#eid=98207.

2 Dioscorides *De Materia Medica* V. 80. For all references to Dioscorides *Materia Medica,* English translation, see Beck 2005.

3 Dioscorides *De Materia Medica* V. 103.

4 Dioscorides *De Materia Medica* V. 82.

5 Dioscorides *De Materia Medica* V. 86.

6 Nicander *Alexipharmaka* 51.

7 http://stephanus.tlg.uci.edu/lsj/#eid=34804; Dioscorides De *Materia Medica* V. 86.

8 *LEXICON ALCHEMECUM* 2.15.

9 http://stephanus.tlg.uci.edu/lsj/#context=lsj&eid=53217.

10 Strabo *Geography* 3.2.8.

the gold is smelted and refined by means of a sort of styptic earth the residuum thereof is called electrum; and again when this electrum which contains silver and gold is smelted the silver is burned away while the gold remains". Slag is also referred to as *kivdos*; [11] *kivdones* [12] are the miners/smelters or those who operated furnaces in the vicinity of the mines.

As to the origin of the word scoria, it derives from scor (σκώρ)[13] meaning *dung, ordure*. It is interesting to speculate that antiquity saw parallels between the fluid "solid" waste (a figurative "excrement") that comes out of a furnace in operation and *that* from cattle. Slag "cakes" with their characteristic ripple-like surface texture, either forms at the base of a furnace, as is the case in bloomery furnaces (Fig. 1) or are tapped and allowed to solidify outside the furnace. Either way they bear an uncanny resemblance to cattle dung, encountered in open fields. These visual similarities would almost certainly not have gone unnoticed.

In this short paper, part of a longer study, currently underway, I suggest that it is important to understand how the ancient world perceived the slags (and not only) they made by paying closer attention to the names they gave to them. Perception and practice are firmly intertwined; by not paying sufficient attention to the former, we potentially run the risk of misinterpreting the latter.

As mentioned earlier, references to metallurgical slag in the ancient texts are relatively few. When they do appear, they are found in pharmacological rather than metallurgical manuals like the afore mentioned Dioscorides (*De Materia Medica*, Book 5) and Galen's (*On Simple Drugs*) (see below) to mention only two. There are many more which can be accessed from the *Thesaurus Linguae Graecae* digital data base. These manuals usually focus on the pharmacological preparations, the mixture of ingredients and how they were put together; however, they provide little background information regarding the method of preparation of the ingredients *per se*.

Galen's discussion of *pharmaka metallica (*medicines which come from the mines) is particularly illuminating.[14] He divides the group in three sub-

Fig. 1. Fragment of bloomery slag; longest axis= 15cm.

categories a. those which are *aftophyei*, natural ones, b. those that derive from the furnaces (*ek kaminou*) whether melting or smelting and he certainly includes scorias (of iron and silver) in that group, and c. the *skevazomena*,[15] those manufactured in a *skevos* (container/utensil) and at ambient temperatures. The third group included the lead-based *psimythion (*near-pure lead carbonate);[16] the copper-based *ios xystos* [17] and zinc-based *psorikon.*[18] *Psimythion* and *ios xystos* became in later periods the renowned white lead (basic lead carbonate or hydrocerussite) and green pigment verdigris respectively, used by artists till modern times.[19] But as indicated already, both were synthesized by the 4th c BCE and were used in pharmacological applications.

In this paper I will argue that the scoria of the three metals iron, lead, silver, are very specific materials. How they were perceived must have had a direct effect on the names they were given. I suggest that they may have acquired their name because either they were metallurgical slag, *per se,* or because they "resembled" metallurgical slag both in "looks" and "origin" (high-temperature reactions). In the first case, they could have been a recyclable material; in the

11 Pollux *Onomasticon Book* 7.99.25.

12 Photius *Lexicon.*

13 http://stephanus.tlg.uci.edu/lsj/#context=lsj&eid=98205

14 On the simple drugs that are extracted from a mine, or a metallurgical furnace see Galen *De simplicium medicamentorum temperamentis et facultatibus,* Kuhn 1821-1833, xii.208.17.

15 On the simple drugs that are *skevazomena*, see Galen *De simplicium medicamentorum temperamentis et facultatibus*, Kuhn 1821–1833, xii.237.14. Galen adds another one, fykos, orchil, the dye produced from fermentation of lichens.

16 Theophrastus *On Stones* 56; Dioscorides *De Materia Medica* V. 88; Photos-Jones *et al.* 2020.

17 Theophrastus *On Stones* 57; Dioscorides *De Materia Medica* V. 79

18 Dioscorides *De Materia Medica* V. 99; Galen *De simplicium medicamentorum temperamentis et facultatibus*, Kuhn 1821-1833, xii.244.17.

19 Gettens *et al* 1967

second case, they could have been made for another industry e.g., glass, glazes, or pharmaceuticals. Emphasizing "intent" may appear pedantic, but in this case, it helps us draw firmly the line between different industries. It also emphasizes the need to look for such materials out with the texts and in the archaeological record.

Scoria and the difference between slag and dross

The Liddell-Scott-Jones (LSJ) dictionary defines *scoria (σκωρία)* as "dross of metal, slag". The use of the two words "dross" and "slag" is of interest here. First, slag is defined as *"a by-product formed in smelting, welding, and other metallurgical and combustion processes from impurities in the metals or ores being treated"*.[20] Modern slag *"consists mostly of mixed oxides of elements such as silicon, sulphur, phosphorus, and aluminum; ash; and products formed in their reactions with furnace linings and fluxing substances"*. As to where slag appears in the furnace*: "it floats on the surface of the molten metal, protecting it from oxidation by the atmosphere and keeping it clean"*.

Slag is not a waste, but a by-product and Greeks aligned their definition of scoria with that provided above: to obtain the "clean" metal, scoria had to be "shaken off" the surface of the metal (*The scoria (of lead) is to be found together with molybdos (lead) and once it is removed, one ends up with the body (metal)*).[21]

Although dross is often assumed to be an antiquated way of denoting slag, in fact "dross of metal" and "slag" can be two different types of materials (metal oxides *vs* silicates) arising from different processes. Dross is defined as a mass of *solid* impurities floating on molten metal and forming on low melting metals such as tin, lead, zinc or aluminium. In antiquity, as today, the method of removing dross from a pot with molten metal consisted of scraping it with a rake or pulling it with a hook. In antiquity, the "*solid*" dross would have been juxtaposed with the "*liquid*" slag and as such may have been perceived differently. "*Solid*" and "*liquid*" refer to the state that the materials were in as they were forming and not after they cooled, at which point they would both be solid. "*Solid*" would involve the solidity of a *lithos* while a "*liquid*" would invoke the fluidity of a slag. As I will try to explain below, this difference becomes important when juxtaposing *scoria argyrou* with *lithargyros* (the *lithos* of silver)[22], another ingredient in pharmacological preparations.

Scoria of iron

Apart from scoria, iron slag was called the *tryx* of sideros*; tryx* is the name given to wine lees or the sediment (spent yeast and solids) that accumulates at the bottom of the fermentation vessels in wine making. Few sources apply the word *tryx* to iron slag.[23]

In his *Metereologica*[24] Aristotle describes iron scoria as the product of *apokatharsis* (purging/cleansing) of iron (*αποκαθαίρεται κάτω η σκωρία*) which falls "below". Although he does not use the word *tryx,* the implication here is that the scoria of iron sinks to the bottom (*κάτω*) of the furnace just like *tryx*, the wine lees, does in a fermentation vessel; this is indeed a correct appraisal of what happens to the bloomery slag which forms around the bloom (*stomoma)* and drips onto the furnace floor in the course of iron smelting. If this assessment is correct, then *tryx* would be of an iron silicate ($FeO/FeO\text{-}SiO_2$) composition and would resemble the bloomery slag shown in Fig. 1. Elsewhere, Pollux [25] defines *scoria* as the *katharma* of the *sideritis ge,* i.e., the iron-rich earth, rather than of *sideros, per se*, which would, metallurgically speaking, be more accurate.

Scoria molybdou

Dioscorides[26] notes that the best *scoria molybdou* is the one that looks like *psimythio*n ($PbCO_3$) (i.e., it is whitish); "*it is dense and difficult to break, it does not look like lead, having instead a quince-yellow colour it is glassy looking*". From this description, the "slag of lead" discussed here cannot be a smelting slag, which is typically black and glassy and consists largely of (Ca/Fe) aluminium silicates, with Pb partitioning between a metallic phase, as Pb metal, and to a smaller extent within the slag.

A *psimythion*-like material, white or near white, which is also glassy and contains Pb alludes to a lead silicate ($PbSiO_3$), deriving from the mixing of PbO with a source of silica. The addition of varying amounts of PbO in silica to produce lead glass has been known

20 https://www.britannica.com/technology/slag

21 Zosimus *Alchemist 2.165 'η σκωρία μετά του μολύβδου ευρίσκεται και αποτινάξαντες την σκωρίαν, έχουσι το σώμα.*

22 Dioscorides *Materia Medica* V. 87.

23 See footnote 6; http://stephanus.tlg.uci.edu/lsj/#context=lsj&eid=108650

24 Aristotle *Metereologica* 383b.

25 Pollux *Onomasticon* Book 7.99.22.

26 Dioscorides *Materia Medica* V.88.

in the Near East for a considerable length of time and lead is known to be present in opaque white glass from the Hellenistic and Roman periods.[27] The white glass decoration of the British Museum's Portland Vase has been found to contain 12% PbO.[28] As a working hypothesis, I put forward the suggestion that the *scoria molybdou* may have been a lead silicate.

Regarding the use of *scoria molybdou* in the medical texts, it was purported to heal ulcers of the nose when mixed with wine and oil;[29] elsewhere it was used having been ground with wine against herpes with blisters.[30]

Scoria argyrou

Dioscorides tells us that the slag of silver is called *elkysma* (*ἡ δὲ τοῦ ἀργύρου σκωρία καλεῖται ἕλκυσμα*).[31] *Elkysma* refers to something that is drawn with a hook. This means that a solid mass (the *elkysma* itself) formed on the surface of the molten metal, and which was subsequently drawn off with a hook. This solid may have been dross and in that case consisted mainly of PbO and other impurities within the silver. The use of a hook to remove the scoria argyrou is rather problematic here because it contradicts the word scoria, assuming the latter did refer to a fluid mass, as I have argued above. If scoria argyrou was indeed PbO (litharge) there is another *pharmakon metallicon* which has also been considered to be PbO, i.e., *lithargyros.* So, which was the "real" PbO? There is a need to be able to differentiate (from the point of view of composition as well as method of preparation) between the two, i.e., *scoria argyrou (elkysma)* and *lithargyros.*

Contrary to *elkysma*, the *lithargyros* of Dioscorides derived directly from the heating at high temperatures *(pyrossis)* of *molybditis ammos* (sand-sized ore rich in lead)[32] and not from silver-rich lead. Since the 1980s, the archaeometallurgical literature regarding Laurion has consistently focused on the production of silver via cupellation of argentiferous lead metal to lead oxide with the subsequent separation of silver from the PbO forming at the bottom of a crucible.[33] Other researchers[34] who were aware of the use of "litharge" in pharmacological applications also suggested that in Laurion, smelting of silver-bearing lead ores took place with subsequent production of litharge for the recovery of silver; they also noted that samples of the litharge are largely absent because they were either recycled (back into the metallurgical furnaces) or sold as pharmaceuticals. But there is no mention of *elkysma.*

PbO exists in two forms, litharge and massicot, of tetragonal and orthorhombic lattice structure respectively. Both are found naturally and have also been synthesized since antiquity, as revealed in the analyses of the Laurion "litharges" by the above authors and in our own work.[35] Litharge forms from the oxidation of molten lead that is stirred, air cooled and ground to form yellow powder; on the other hand, massicot is formed by the heating of lead carbonate to 300°C.[36] As mentioned earlier, there is a real need to establish how antiquity perceived the difference between *elkysma* and *lithargyros,* since this difference is not presently clear.

Conclusion

I argue that it is important to establish conclusively the nature and method of manufacture of many pyrometallurgically-derived compounds which were used as ingredients in pharmacological preparations, and which are presently little understood. Even less understood is why they were deemed efficacious. I focus on three: the scoria of lead, iron, and silver. I argue that scoria denoted a certain fluidity because of the way it emerged out of the furnace/hearth (perhaps with a rippled surface) as well as in a rounded shape (perhaps a cake). Scoria was a by-product of smelting, but a scoria-looking material could be the main product of other high-temperature practices like glassmaking or glaze-making. Compositionally it must have had a silicate base.

The *scoria siderou*, I suggest, may have been equivalent to the bloomery slag on account of the analogy drawn with *tryx* (its other name), in both cases a sediment accumulating at the bottom of either the furnace or the fermentation tank. This material would have been derived from bloomery iron making and could have been added to pharmacological preparations.

27 Charleston 1960; TAIT 1991.

28 Freestone 1990.

29 Oribasius *Med. Eclogae* 11.1.

30 Oribasius *Med. Eclogae* 104.1.

31 Dioscorides *Materia Medica* V.86.

32 Dioscorides *Materia Medica* V.87.

33 Conophagos 1980.

34 Rehren *et al.* 1999.

35 Photos-Jones and Jones 1994.

36 Morales *et al.* 1985; Gliozzo and Ionescu 2022.

Scoria siderou together with all other scorias were thought to be "drying". The former was applied, amongst others, to a "pussing ear" (*pyorrounta ota*) (*πυορροούντα ώτα*) after it had been finely ground with oxos (old wine) heated to a high temperature. It is difficult to understand how finely ground slag mixed with "vinegar" could be effective against a discharging "wound". However, some insight, albeit tenuous, may be gained from early 20th c research[37] which has shown that silica gel (non-crystalline SiO_2) is an effective catalyst for the conversion of acetic acid to acetone. Acetone is a very effective antibacterial agent and is used for the sterilisation of surfaces, particularly instruments, prior to surgery[38] but is not to be applied onto the skin. Acetone has a fruity smell and if produced in the ancient workshop it would have been "picked up", meaning it would have left an impression as distinct and different. However, all these suggestions are speculative, and only justified by our curiosity to explain this transition of slags from metallurgy to pharmacology.

Regarding *scoria molybdou,* I suggest that if it were to warrant its name as a slag, it should have been a viscous material, and a Pb-silicate on account of needing to be white and glassy. It is reported that Pb-silicate is made from Pb-acetate with a sodium silicate source.[39]

Regarding the *scoria argyrou*, also called *elkysma*, I highlighted several issues. Assuming the *elkysma* was a solid, *lithos*-like PbO based material, I am presently unable to discern the difference between *elkysma* and *lithargyros* despite the texts making it clear that these were two different materials.

The medical texts of antiquity explain that, individually and/or combined, Fe and Pb metal oxides and silicate-rich material (slag) arising from metal-making practices found useful applications in pharmacology. This may serve as an incentive for us to begin to re-evaluate our appraisal of (some) archaeometallurgical slag as a waste material and perhaps try to locate pharmacological slags in the archaeological record.

Acknowldgements

This work continues from an earlier award (Seed Award in the Humanities and Social Sciences) (201676/Z/16/Z) by the Wellcome Trust to the present author; the aim of the project was and contiues to be about setting minerals known in antiquity for their medicinal applications into a modern pharmacological context.

37 Mitchell and Reid 1931.

38 Drew 1977.

39 https://cameo.mfa.org/wiki/Lead-silicate

Bibliography

L. Beck, *Pedanius Dioscorides of Anazarbus De Materia Medica*: Translated and edited by Lily Y. Beck. 2005 Altertumswissenschaftliche Texte und Studien 38. Hildesheim.

C.E. Conophagos, *Le Laurium antique et la technique grecque de la production de l'argent,* Athens, 1980.

Galen *De simplicium medicamentorum temperamentis et facultatibus*, KUHN 1821-1833. Volume 12 https://www.biusante.parisdescartes.fr/histoire/medica/resultats/index.php?do=page&cote=45674x12&p=5.

E. Gliozzo and C. Ionescu, "Pigments - Lead-based whites, reds, yellows and oranges and their alteration phases", *Archaeological and Anthropological Sciences* 14 (2022): 17. https://doi.org/10.1007/s12520-021-01407-z.

R.C. Drews, "Acetone sterilization in ophthalmic surgery", *Ann. Ophthalmol.* 9, 6 (1977), 781-784. PMID: 41031946.

R.J. Charleston, "Lead in Glass", *Archaeometry* 3,1 (1960), 1-4. https://doi.org/10.1111/j.1475-4754.1960.tb00508.x.

I. Freestone, "Laboratory Studies of the Portland vase", *J. Glass Studies* 32, (1990), 103-107. https://www.jstor.org/stable/24188034.

R.J. Gettens, H. Kuhn, and W.t. Chase "Lead white". *Studies in Conservation* 12:4, (1967), 125-139.

J.A. Mitchell and E.E. Reid, "The decomposition of acetic acid in the presence of silica gel", *J. Am. Chem. Soc.* 53, 1 (1931), 338–343. https://doi.org/10.1021/ja01352a050.

J. Morales, J.l. Tirado, M. Macias, A. Ortega, "Influence of crystallinity on the kinetics of the litharge-massicot phase transition", *Reactivity of Solids* 1, 1 (1985), 43-55. https://doi.org/10.1016/0168-7336(85)80006-1.

E. Photos and J.E. Jones, "The building and industrial remains at Agrileza, Laurion (fourth century BC) and their contribution to the workings at the site", *Annual of the British School at Athens* 89 (1994), 307-358.

E. Photos-Jones, P. Bots, E. Oikonomou, A. Hamilton, C.W. Knapp, "On metal and 'spoiled' wine: analysing psimythion (synthetic cerussite) pellets (5th-3rd centuries BCE) and hypothesising gas-metal reactions over a fermenting liquid within a Greek pot", *Archaeological and Anthropological Sciences* 12, 10 (2020) https://doi.org/10.1007/s12520-020-01184-1.

Th. Rehren, D. Vanhove, H. Mussche, M. Oikonomakou, "Litharge from Laurion. A medical and metallurgical commodity from South Attika", *L'antiquité classique* 68 (1999), 299-308.

H. Tait (ed.), *Five thousand years of Glass,* London, 1991.

9. METAL ARTEFACTS AND JEWELLERY IN THE STRYMON RIVER PLAIN: TWO ARCHAIC "THRACIAN" GRAVES FROM THE AMPHIPOLIS REGION

Ioannis Chalazonitis and Dimitria Malamidou

Introduction

Other colleagues in this volume have shed considerable light on the complex *chaîne opératoire* of mining, ore processing and manufacturing of metal objects in early Mediterranean communities. This paper complements these proceedings with a divergent viewpoint: one that does not focus on the creation process of metal goods but, instead, on the end of their life and, specifically, their final deposition in funerary assemblages.

The two graves serving as our case studies for this paper were excavated by the Serres Ephorate of Antiquities and examined for publication in the context of the research project titled "Legacy of the Riverland Heroes", kindly hosted by the CReA-Patrimoine and funded by the FNS-FNRS.[1] The project seeks to explore the influence of Early Iron Age and archaic elite burials as axes of social cohesion among communities of the Strymon river plain (for example as important territorial landmarks, or loci of ritual activity). To achieve our overarching research goals, it became imperative to familiarise ourselves with the widest possible spectrum of funerary ritual in the region for two primary reasons: firstly, to establish the characteristics of the regional "baseline" funerary *habitus* against which "exceptional" elite burials would be better outlined and, secondly, to better understand how these communities engaged with their dead. Within this framework, the study of grave good assemblages, and particularly metal offerings was paramount: the relatively high material value of some of those artefacts (when compared to other grave goods), in addition to the specialised technological know-how required for their creation, highlight them as potential indicators of elite social status and as potential semantic features of signalled social identities.

The objective of this paper is twofold. Firstly, we aim to demonstrate the relative ubiquity of specific types of metal grave goods in archaic Thracian burial assemblages of the Strymon river plain: this will involve the presentation of our two case studies, followed by a brief overview of other known funerary contexts of the region. Secondly, we seek to determine whether the two studied graves are sufficiently divergent from the regional baseline to justify them being ascribed to local "elites": preeminent and wealthy individuals, with a privileged access to local and imported resources, and with relatively high prestige and influence in and beyond the local community. As part of this second axis of investigation, we also attempt to interpret the potential semantic significance of specific types of metal grave goods, with particular emphasis placed on their role as representative of "desired" or "idealised" social roles and identities. To achieve this latter goal, we will seek out and attempt to interpret repeated patterns of grave good deposition in the region.

The graves

Grave A

The grave was discovered in October 1999, during rescue excavations conducted by the Serres Ephorate of Antiquities near "Hill 133', in the region north of

1 We wish to warmly thank the community of the CReA-Patrimoine for their support and feedback throughout the duration of the project in question. Special thanks go to Professor Athéna Tsingarida and Professor Didier Viviers for their advice and unwavering encouragement; Dr. Vivi Saripanidi for inspiring the project and for her invaluable and continued critical feedback; and, last but not least, Dr. Nerantzis Nerantzis for his kind invitation to participate in these conference proceedings. Among our colleagues in Greece, we warmly thank the personnel of the Ephorate of Serres and of the Ephorate of Kavala (Mrs. Stavroula Dadaki and Mrs. Pinelopi Malama in particular) for their hospitality during the study of the material, as well as Dr. Chaido Koukouli-Chryssanthaki for her generosity, kindness, and peerless insight into the sites of the region. Finally, we gratefully acknowledge the financial support of the FRS-FNRS, as well as the generous additional funding awarded by the Union Académique Internationale (Brepols Prize for Early Career Scholars), without which the project could never have been realised.

1

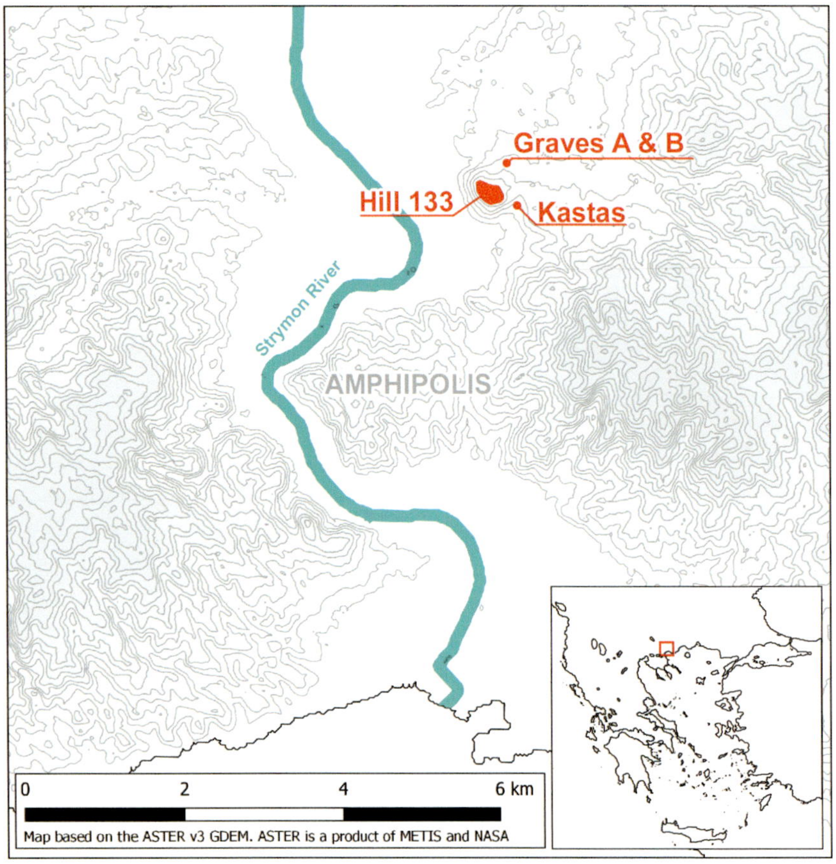

Fig 1. Map of the area near Hill 133. Map is based on the ASTER 3 GDEM. ASTER is a product of METIS and NASA.

Fig 2. Pottery assemblage from Grave A (© Ephorate of Antiquities of Serres & Greek Ministry of Culture).

2

Amphipolis (Fig. 1).[2] It was a cist grave, built with four large schist slabs, oriented NE-SW, and containing a single, undisturbed inhumation. Unfortunately, the poor taphonomic conditions contributed to the loss of all skeletal material except for part of the skull, which was discovered at the NE end of the grave.

The pottery from the grave will not be our primary focus of investigation within the framework of this paper, but it does provide crucial dating information for the burial assemblage, and therefore merits a brief

2 Malamidou 1999, 695; 2006, 189-90.

discussion. The grave contained four clay vessels: A locally produced trefoil oenochoe with horizontal bands on the belly and a wavy-line motif on the neck and handle,[3] a large "Thasian" cup, a small Corinthian kotyliske and a spherical / globular Corinthian aryballos (Fig. 2). The "Thasian" kylix, decorated with painted concentric circles and sets of vertical lines, is of note, as it belongs to the so-called "cup with notches" type (*"coupe à encoches")*, originating in Paros but later produced in Phari, in Thasos, and possibly in other workshops of the Thasian region.[4] Typically, the sides of such cups found in the Phari workshop are each divided into two large zones by three groups of six or seven vertical lines; each zone contains a Group of concentric circles (usually four). Our example belongs to a less common type, with only three concentric circles and five vertical lines per respective group.[5]

The two Corinthian vessels from Grave A form part of a larger corpus of such imported vessels in the region: Corinthian wares (usually exaleiptra, aryballoi and kotyle cups) are present in many late archaic burials of Aegean Thrace.[6] In this instance, the lower part of the kotyliske bears the succession of red-brown and black bands often encountered in archaic miniature vessels of this type, while its lip is decorated with the so-called "Antiparos Pattern", i.e. two series of opposed and alternating stepped triangles.[7] The aryballos bears a quatrefoil motif, which is, unfortunately, very poorly preserved.[8] As a whole, the pottery dates the burial assemblage with certainty to the second half of the 6th century;[9] the Thasian cup, in particular, indicates a date towards the final quarter of the 6th century (or the earliest years of the 5th) as most likely.[10]

The metal finds from the grave reinforce this dating. We must first mention the pair of bronze bracelets - the only bronze objects discovered in the grave - found clipped together, much like a chain's interlocked links (Fig. 3). They belong to a type that is relatively common during the second half of the 6th century and the early years of the 5th century BCE, in Epirus, Macedonia and western Thrace. Similar bracelets are also encountered further north in the Balkans, including but not limited to sites of Albania and Northern Macedonia.[11] They are open rings, with a circular cross-section; their endpoints are shaped like reverse rounded pyramids with abutting bases. Thin relief bands mark the boundaries between the endpoints and the main body of the bracelets. A shallowly incised helical groove decorates the 1-2 cm closest to the endpoints.

Such bracelets were typically worn on the wrists, forearms, or upper arms (see below, Grave 2); in this instance, however, the loss of the skeletal remains does not allow us to determine whether the bracelets were, in fact, worn by the deceased. The fact that the two rings were clipped together might have prevented their use as personal ornaments, and they may have been simply deposited in the grave beside the body. It is worth noting that another instance of such interlocking bracelets, dating to the final quarter of the 6th century BCE, is known to us from the archaic necropolis of Agia Paraskevi, south-east of Thessaloniki. Sadly, the grave in question had been looted, and the original placement of these bracelets is, likewise, unknown.[12]

Grave also contained three iron objects: an iron spearhead, a curved iron knife and fragments of an iron wire belt (Fig. 4). As is typical for iron finds of

3 The oenochoe belongs to a variant of the so-called "Wave-line pottery" style: a style which is often encountered in the region, and which has been linked to the adoption of Ionian influences. See Chavela 2013, 201-4; Perron 2013a, 133-42; Perron 2013b, 88-90, 223-4, 248-50, 304-9 with previous bibliography.

4 The type has received its name after the three small, incised "notches" borne by some of these cups between the attachment points of at least one of their handles. See Dugas 1928, 192 for the first use of the term; see also Blondé, Péristéri, & Perreault 1992, 24-31 for the publication of several such cups from Phari.

5 Blondé, Péristéri, & Perreault 1992, 24.

6 See below, "Metal Grave Goods in other Burial Grounds" for selected examples of such pottery from other cemeteries of the Strymon river plain.

7 For the first identification of the "Antiparos Pattern", see Payne 1931, 334. For some typological parallels of the kotyliske see Boardman & Hayes 1973, 13-4; Risser 2001, 70, no. 204; Neeft 2006, 97 with previous bibliography.

8 The motif can readily be identified as Payne's variant E (1931, 147); however, its poor state of preservation does not allow a confident attribution to a particular type. Barely visible traces of a small row of dots above the center of the quatrefoil, on the shoulder of the vase, would point us towards Hayes' Type II, dated to the transitional Phase between Middle and Late Corinthian (Boardman & Hayes 1973, 22 and 30). Similar aryballoi from the necropolis of Sindos have been dated to the third quarter of the 6th century (Saripanidi 2016, 54-57 with previous bibliography). For the region of the mouth of the Strymon, see also parallels from settlement contexts in Argilos, recently published by Neeft 2020, 21-2.

9 All dates of archaeological material henceforth are BCE, unless indicated otherwise.

10 Blondé, Péristéri, & Perreault 1992, 30.

11 See Misailidou-Despotidou 2011, 176-9 with exhaustive previous bibliography.

12 Misailidou-Despotidou 2011, Cat. No. 559-560.

3 

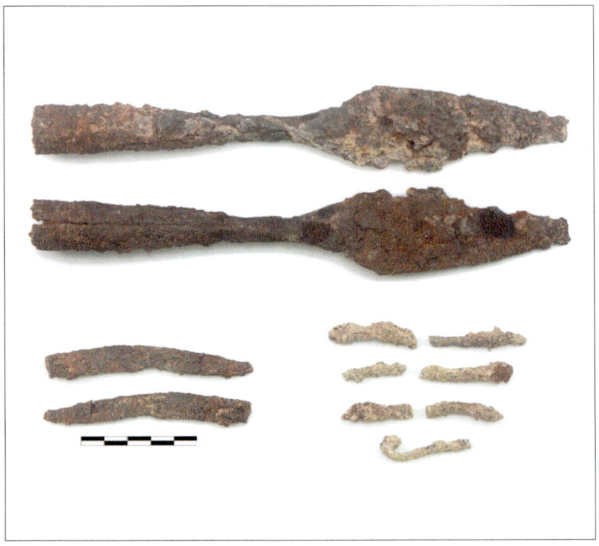 4

Fig 3. Bronze bracelets from Grave A (© Ephorate of Antiquities of Serres & Greek Ministry of Culture).

Fig 4. Iron spearhead (both sides), knife (both sides), and belt from Grave A (© Ephorate of Antiquities of Serres & Greek Ministry of Culture).

the period in the region, their state of preservation is relatively poor: their surface is heavily oxidised, and parts of the blades and belt are missing. This renders a typological study of these finds relatively difficult; however, it is possible to note the following:

The spearhead is 23.5 cm long, including its socket, which accounts for almost half of its total length. The "neck" of the spearhead (i.e. the point where the blade meets the socket) is particularly thin, and widens relatively sharply to form the angular "shoulders" of the blade itself. A midrib may have been present but, if so, it has been lost to oxidation. However, the socket still preserves traces of the wooden haft of the spear; this confirms that the spearhead originally formed part of a potentially functional weapon.

The typology of the weapon recalls similar examples from other funerary contexts in the northern Aegean. Indicatively, we note the well-published finds from Sindos, which present us with weapons similar in chronology, size, and form.[13] Baitinger's study of the types dedicated in Olympia is also noteworthy: our spearhead would belong to his A5 type, i.e. one of the heaviest types of throwing-spear found in the sanctuary.[14] In the cultural context of Grave A, it should probably be interpreted as a skirmisher weapon or hunting spear.

The knife found near the waist of the deceased is of a recurve type, not so common during the 6th century BCE, but with a very long lifetime, dating at least as far back as the Late Bronze Age in the north-eastern Aegean (and significantly earlier in other Aegean regions).[15] A very close parallel from eastern Macedonia comes from the Early Iron Age tumuli excavated near the Industrial Zone of Drama.[16] Typical 6th-century knives in the northern Aegean present us with a simpler profile: either a straight spine, or a simple, slight curve.[17]

The final iron object recovered from Grave A was an iron wire belt with spiral endpoints, found in several fragments. This is an ornament and a piece of personal kit that is found in a relatively small number

13 Despoini 2016, 307-13, particularly Cat. Nos. 637 and 642. See also Snodgrass 1964, 127-8 with commentary for what we believe to be the most similar variant.

14 Baitinger 2001, 46-7, Pl. 23-4, particularly Cat. Nos. 636, 639 and 649.

15 Koukouli-Chrysanthaki 1992, 405, type EV1a; see also 587-588 with previous bibliography.

16 Museum of Drama, Cat. No 389. The knife is dated to the Early Iron Age, and was found in tumulus A. Its exact dating is uncertain, but the most recent grave from that specific tumulus dates to the 7th century. For the tumulus burials in general see Koukouli-Chrysanthaki 1979, 333-4; 1993, 685-6.

17 See e.g. the knives from Sindos Despoini 2016, Figs. 190-215; ongoing research by the authors in the site of Tragilos indicates that the same knife typology is prevalent there as well. Note in particular Despoini 2016, Fig. 200 and 210 for potential parallels.

Fig 5. Pottery assemblage from Grave B (© Ephorate of Antiquities of Serres & Greek Ministry of Culture).

of 6th-century graves in northern Greece: examples have been discovered in Kastas, in the tumulus of Alepotrypa, in Tragilos, in Gazoros, in Eion and in other sites of the Strymon river plain.[18] Similar belts, and variants, have also been discovered further west, in sites such as Vergina, Archontiko and possibly Mavropigi (the examples found there being large iron hoops rather than torc-like in shape).[19] To the best of our knowledge, belts of this type appear almost exclusively in male burials and, as is the case here, are usually accompanied by weaponry, such as spearheads, swords and, most commonly, knives. Traces of leather or fabric discovered on some of them might indicate that they had originally been wrapped in a decorated "sleeve", to protect the iron and avoid oxidation stains on their users' clothes: a silver instance of such a belt discovered in Trebeništa shows a broad band with intricate patterns on its back, which may reflect similar decoration in the "sleeves" of more common, iron examples.[20] Grave A presents us with a variant that is rare in eastern Macedonia / western Thrace, where the spirals are relatively small and angled against the rest of the wire.[21] It is worth noting that the endpoints are somewhat reminiscent of deformed nail heads, which led to them being originally misidentified as nails.

Grave B

The second grave was discovered, same as Grave A, near Hill 133, in October 2003 in the course of agricultural works, and excavated in a rescue excavation;; it is thought that the two graves formed part of a larger, burial ground that included several graves of the same period.[22] It is very likely that such a burial area would have existed in parallel with (or as an extension of) the well-known archaic necropolis of Kastas, which has also yielded graves dating to the late 6th century BCE, and which lies to the immediate east of the nearby "Hill 133'.[23]

Grave B was a slab-built cist grave, but its state of preservation was notably worse than that of Grave A. It was not looted or artificially disturbed, but the eastern (long) slab of the grave had shattered under the pressure of the surrounding earth, and its fragments had been pushed inwards, partially displacing the poorly preserved skeletal remains. The grave goods themselves remained in relatively good condition.

18 See below, "Metal Grave Goods in other Burial Grounds".

19 Vergina: Kontogoulidou 2007, 146, fig.7; Kakamanoudis 2018, 108; Archontiko: Chrysostomou & Chrysostomou 2001, 484, Fig. 12; 2006, 705-6; Mavropigi: Karamitrou-Mentesidi 2013, 78-9.

20 Ardjanliev *et al.* (eds) 2018, Cat. No. 108.

21 Another belt of this type has been discovered in ancient Tragilos, in a funerary urn containing a secondary cremation (Amphipolis Museum, Cat. No. Δ16976). The silver example from Trebeništa (see above, n. 20) also has spiral endpoints that are similar to this variant.

22 Malamidou 1999, 695; 2006, 188-90.

23 See below, "Metal Grave Goods in other Burial Grounds"; Lazaridis 1964 - 1982.

Fig 6. Fragments of silver mouth-covering and pair of gilded silver earrings from Grave B (© Ephorate of Antiquities of Serres & Greek Ministry of Culture).

The pottery recovered from the grave does not include any Corinthian vessels. Instead, it comprises an undecorated oenochoe (in all probability a local product), accompanied by a "Thasian" kylix, near-identical to the one discovered in Grave A (Fig. 5). Once again, each side of the vessel is divided into two parts by three sets of five grouped vertical lines; at the centres of the resulting fields, we find sets of three concentric circles. The only differences between this cup and the one from Grave A is a more concave foot, and a reserved band, running the circumference of the vessel, halfway down its otherwise uniformly painted lower part.

Grave B may have yielded fewer pottery vessels than Grave A, but its corpus of metal grave goods is considerably richer. The poor preservation of the skeletal remains does not allow a forensic identification of the sex of the deceased, but the choice of metal grave goods hints towards a woman's grave:
On the neck area of the deceased, fragments of silver foil with decoration in *repoussé* were discovered. The foil is less than a third of a millimeter thick and extremely fragile: the original object has fragmented into more than twenty larger fragments (of which a selection with visible decoration is shown in [Fig. 6]), and a multitude of smaller ones. It is impossible to fully reconstruct the original object or determine its function with certainty. However, given its placement in the grave, it is likely to have been a silver mouth-covering. These ornaments, typically made of gold or silver foil, were sewn onto a cloth backing and placed over the mouth of the deceased, not unlike a death-mask[24]. In some instances, they were accompanied by similar coverings for the forehead, eyes and/or nose, and their decoration ranged from simple geometric motifs to miniature pictorial scenes. In this instance, the curved and straight lines that can still be identified on some of the fragments argue for a typical rhomboid design with a central motif: possibly a rosette, not unlike those decorating some of the mouth-coverings discovered in Sindos.[25]

In addition to the mouth-covering, a pair of silver earrings were found near the head (Fig. 6). They are made of thin, gilded silver wire, which was folded into two parallel strands, and wound into two complete spirals. At their endpoints opposite to this primary fold, the wires were bent into an elegant, elongated cardioid pattern, similar to motifs found in dress-pins of the period.[26] Four silver beads per earring, also gilded, were placed on nodal points of the endpoint

24 A very fragmentary iron wire or rod, found under the skull, may have originally been part of a dress-pin, holding together this cloth backing or the shroud. This object is very poorly preserved, and its interpretation is purely speculative.

25 Despoini 2016, 22-31, 417-9 with previous bibliography; also Saripanidi 2017, 106-10..

26 Misailidou-Despotidou 2011, Cat. No. 250-98; Despoini 2016, 166-76.

pattern, to complete the design. The earrings are mirrored versions of each other, clearly forming a purpose-made pair, and their state of preservation is excellent, except for a deformation and break sustained by the first *c.* 20mm of one of the two.[27] Earrings of this type have been encountered in other burial sites of the northern Aegean, where they have been sometimes mistaken for large hair-coils or bracelets.[28] In the Strymon region, we find two pairs of near-exact parallels in the archaic necropolis of Tragilos (the only difference being the use of fewer silver beads per earring),[29] but other examples are also known from sites further to the west. The necropolis of Agia Paraskevi has yielded a pair of such earrings made of bronze,[30] and another silver pair has been discovered in Epanomi.[31] The latter example is the only known case where the significantly smaller bronze rings, used to suspend the silver spirals from the ears of their owner, had been preserved.

Grave B also yielded two bronze bracelets. They are, in all respects, identical to the ones discovered in Grave A, including their overall dimensions and decoration (Fig. 7).[32] However, unlike Grave A, they were not clipped together, but worn by the deceased: one was worn on the lower left arm, while the other was worn on the upper right arm. Assuming that our identification of Grave A as a male burial and Grave B as a female burial is correct, this does raise questions regarding the significance of (and value ascribed to) such personal ornaments, particularly with regards to gender. In burial grounds further west, bracelets of this type are found overwhelmingly

Fig 7. Bronze bracelets and biconical bead from Grave B (© Ephorate of Antiquities of Serres & Greek Ministry of Culture).

in female burials.[33] Whether the difference in how the bracelets were deposited is relevant remains unclear for the time being, but, hopefully, the ongoing study of other grave contexts from the Strymon river plain will provide additional data points and further insight into local practices.

To the immediate left of the neck of the deceased, a large biconical bead was brought to light (Fig. 7). It may originally have been worn as part of a necklace. Such beads present us with a large variety of subtypes and were manufactured using the "lost-wax" technique.[34] In this instance, our bead belongs to subtype ΨΗ-Γβ1 as proposed by Misailidou-Despotidou, with a central hole that is smaller than the overall length of the bead, a relatively linear (straight) conical profile, and ring-like endpoints. Beads of this or related subtypes have been found throughout the Aegean, the Greek mainland, and the southern Balkans (Albania, North Macedonia, Bulgaria); their earliest appearance could be dated back as far as the early 7th century BCE.[35] They are certainly common in the Strymon river plain, with other known examples from Amphipolis, Serres and Tragilos.[36]

Among the finger-bones of the left hand, fragments of a bronze ring were discovered. Bronze rings are relatively common in late archaic burials, and they range in complexity from simple rings with a circular cross-section, to undecorated flat bands (as is the case here) to complex pieces of jewellery with two or

27 This break is somewhat fortuitous in that it allows examination of the inner structure of the wire. Preliminary observations, pending a more detailed metallographic examination of the object, seem to indicate that the wire was created with the strip-twisting technique (see Carroll 1972, 321-3; Oddy 1977, 85-6)..

28 For the problems faced in identifying the function of these objects see Tsimbidou-Auloniti 1989, 322-3; Misailidou-Despotidou 2011, 46-7 with previous bibliography. See also Laffineur 1980, Cat. No. 24 for a pair of such earrings, originally interpreted as bracelets, in the P. Kanellopoulos collection. Notably, the silver earrings from Grave B had originally been interpreted as hair-coils by the excavators and the authors. We wish to warmly thank Dr. A. Touloumtzidou for correcting our error and bringing to our attention the parallels from Epanomi (see below, n. 31), which conclusively demonstrate the original use of such objects as earrings.

29 Kavala Museum, Cat. No. M1606 (three beads), M1615 (three beads), M1627 (two beads), M1628 (four beads).

30 Misailidou-Despotidou 2011, 46-7.

31 Tsimbidou-Auloniti 1989, 322-3.

32 See above, n. 11.

33 See e.g. Agia Paraskevi: Misailidou-Despotidou 2011, 142. Out of 120 graves, yielding a total of 321 bracelets, only three pairs were found in burials that are tentatively identified as male.

34 Misailidou-Despotidou 2011, 38-40.

35 Misailidou-Despotidou 2011, 89-95.

36 Bouzek 1974, 109-12, Fig. 34 ; Koukouli-Chrysanthaki 1983, 139.

Fig 8. Bronze spectacle fibulae from Grave B (© Ephorate of Antiquities of Serres & Greek Ministry of Culture).

four spiral endpoints, arranged symmetrically. In the northern Aegean regions, they appear in both male and female burials; in some burials, multiple rings are worn by the same person.[37]

A fragmentary bronze needle and three spectacle fibulae of significantly different dimensions and weight complete the corpus of metal grave goods from Grave B (Fig. 8). The first one, originally placed on the left shoulder of the deceased, is very large, reaching just under 175mm in total length. The wire has a rhomboid cross-section and forms a secondary figure-of-eight motif (also known as *"Achterschleife"* in previous bibliography)[38] between the two large spirals. The fibula belongs to J. Alexander's "Type Ib", the origins of which date as far back as the 11th century. The type is encountered in many sites of south-east Europe, the Balkans, and the Aegean; its lifetime is extensive, as it remains in use until at least the 5th century.[39] However, during the late 6th and early 5th centuries, it becomes increasingly uncommon in eastern Macedonia: a small number of parallels from nearby Tragilos remain the only other known large-format spectacle fibulae from the Strymon river plain during the late archaic period.[40]

The second spectacle fibula was found to the immediate right of the skull, next to a fragmentary bronze needle. It is likely that, originally, the two objects had been placed next to the body instead of being worn by the deceased. This second fibula is relatively well-preserved, more than four times smaller than the first (only being *c.* 40mm long), but of the same basic type. One difference lies in how the cross-section of the wire is arranged: both wires have a square cross-section, but in the first fibula the wire is twisted by forty-five degrees, resulting in a rhomboid effect. In the second fibula, the wire is kept flat, resulting in a smoother overall impression for the spirals.

The final fibula is very small (its original length is estimated to have been under 35mm) and destroyed in its near entirety: only part of one of the spirals and many smaller fragments have been preserved. It was found away from the body, against the north-east corner of the grave. Its type is impossible to identify with any certainty due to its poor state of preservation, but a sharp curve in one of the smaller fragments might argue in favour of a central figure-of-eight motif. However, one significant difference that sets it

37 For a detailed overview of the development of the extremely varied typology of bronze rings, see Misailidou-Despotidou 2011, 186-98 (particularly 190-2 for the flat band type encountered in Grave B); see also *ibid,* 229-30 for their use as grave goods in both male and female graves.

38 See e.g. Pabst 2011.

39 Alexander 1965, 8-11 and also 20-1 for a site index with previous bibliography; see also Pabst 2011, 211-3 for a study focused on the genesis of the type (which she calls the "Vergina-Type") during the Early Iron Age and Misailidou-Despotidou 2011, 95-7 for an updated site index focusing primarily on Macedonia.

40 Koukouli-Chrysanthaki 1971, 418 and Pl. 415; see also Misailidou-Despotidou 2011, 97 for the few other late archaic parallels.

apart from the other two fibulae is that its wire has a circular cross-section. Whether these differences were stylistically, chronologically, or typologically significant remains unclear. Differences in wire cross-section between different types of spectacle fibulae have not yet been convincingly tied to dating; instead, they may have been linked to the overall size of the fibula.[41]

Metal Grave Goods in other Burial Grounds

The Strymon river plain has yielded several known late archaic burial grounds. Most of these funerary contexts have yet to be published systematically, but excavation reports, preliminary analyses of archaeological material and past synthetical studies can assist us in forming a relatively detailed impression of the established burial practices in our timeframe.

The archaic necropolis most closely associated with our graves is, without a doubt, that of the hill of Kastas, explored during excavations in the 1960s and 1970s.[42] Multiple test trenches have revealed more than eighty individual graves, ranging in date from the Early Iron Age to the Hellenistic Period. At least thirty of those graves can be safely dated to the second half of the 6th and the early 5th century; if one also includes neighbouring graves with no grave goods, this number rises to more than fifty.[43] They are typically cist graves (accompanied by a smaller number of pit graves) and appear to have been organized in loose clusters throughout the hill, instead of being concentrated in a single, extensive cemetery. It is likely that the Graves A and B presented here used to belong to satellite grave clusters of this burial ground.[44]

Excavation reports give us a good overview of the individual grave good assemblages, although in some instances the finds from several graves are presented together.[45] Child and adult graves are both represented in Kastas; the latter typically contain more offerings, but there are instances of both types with no grave goods. When grave goods are present, they almost always include one pottery vessel: Corinthian exaleiptra ("kothons", in Lazaridis' reports) are the most common vases, but a single Protocorinthian aryballos, a few amphoriskoi, two black-figure lekythoi and several kylix cups are also reported. Only two graves contain more than one vessel.[46]

These pottery vessels are almost always accompanied by one or more metal grave goods. If weapons are present, these are typically a single spearhead and / or a knife, sometimes accompanied by an iron wire belt and one or two other objects. In Kastas, such belts can, on occasion, be encountered as the sole metal grave good or accompanied by gold foil mouth-coverings.[47] Graves with iron grave goods, whether weapons or belts, are characterised by relatively small assemblages of one to four objects at most.

However, when weapons and belts are absent, graves in Kastas present us with more variety in the number and the material value of the grave goods. Among such graves, the richest seem to have belonged to women, as is indicated by the presence of earrings, bracelets, and pendants. The richest grave from Kastas, Grave 29, belonged to a woman: it contained four bronze fibulae, a bronze dress-pin, a silver ring, two golden earrings "decorated with the granulation technique", a golden bead, a golden mouth-covering and a Corinthian exaleiptron.[48] It is, in most respects, comparable to our Grave B, with some notable differences (notably the absence of bracelets and the presence of an exaleiptron instead of a drinking cup and oenochoe). Given the above, our Grave A seems to be richer than most comparable graves with weapons in Kastas (the presence of the two bronze bracelets and multiple pottery vessels being particularly notable). Similarly, Grave B seems to compare favourably to the richest Kastas graves.

Further to the north, late 6th-century burial grounds have been discovered near modern-day Agios Christoforos / Nea Zichni and Neos Skopos. In previous bibliography, these two sites have been identified as ancient Gazoros and Vergi, respectively.[49]

41 Misailidou-Despotidou 2011, 97; see also Kilian 1975, 144; Rhomiopoulou & Kilian-Dirlmeier 1989, 117.

42 For the early cemeteries, see Lazaridis 1964, 39; 1965, 49-50; 1971, 50-62; 1973, 50-4; 1974, 63-4; 1975, 69-71; 1976, 94-8; 1977, 42-4; 1978, 57-8; 1979, 79; 1981, 25; 1982, 50-1; for the transformation of the natural hill into a hellenistic tumulus, see Peristeri 2010; 2012; 2013; 2015.

43 See Kakamanoudis 2018, 361-7 for a synthetic overview of the archaeological register, with previous bibliography.

44 Malamidou 1999, 695; Kakamanoudis 2018, 366.

45 Lazaridis 1981, 25.

46 See above, n. 42. The male Grave 15 yielded an amphoriskos and a banded kylix cup (Lazaridis 1975, 71) and the child Grave 28 yielded the only two reported black-figure lekythoi from the cemetery (Lazaridis 1978, 58).

47 See e.g. Kastas Grave 22 (Lazaridis 1976, 95; see also Stamatopoulou 2014, 284-7.).

48 Lazaridis 1979, 79.

49 Gazoros: Samsaris 1976, 130; Vergi: Poulios 2001, 219-240; Valla 2003, 111. See also Zannis 2014, 139-40,

In ancient Gazoros, a relatively large necropolis located to the south-west of the ancient habitat space extends over an area of around nine hectares. It comprises more than 50 excavated graves dating from the late 6th / early 5th century to at least the 3rd century. Most are slab-built cist graves belonging to adults, but at least 13 children and infants were also interred in this area, in pit burials and *enchytrismoi*. Excavation reports have provided us with a detailed view of some assemblages, but a detailed overview of most graves is unavailable. In addition, the destruction of a significant number of the graves by agricultural works and looting only serves to render an investigation of the material more difficult: less than half of the excavated graves were undisturbed.[50]

The exact number of graves dating from the late 6th until the first quarter of the 5th century (inclusive) is unclear: at least 6 known assemblages can be dated to this period, but more are implied in the early reports.[51] For this phase, imported pottery is present (three instances of Corinthian exaleiptra and several Attic lekythoi being particularly notable), but locally-produced pottery is also encountered. Metal finds are present and prominent in all reported early assemblages, and they typically include golden mouth-coverings (in some instances only fragmentarily preserved), bronze bow fibulae, pins, and rings.[52]

Grave 20, excavated in 1995 by V. Poulios and dating to *c.* 475 BCE (almost a generation after our Graves A and B) presents us with high-quality jewellery and rich assemblages that match or arguably surpass any of those found in Kastas. It yielded two Attic lekythoi; a locally produced exaleiptron; two amphoriskoi; a melted glass vase; a pair of silver earrings; two silver rings, one of which bore a seal; nineteen bronze bow fibulae; two bronze rings; two bronze bracelets; bronze and iron pins, possibly linked to fragments of two ivory spectacle fibulae; and seventeen bone and glass necklace beads.[53] While this grave is considerably richer than any we've discussed so far (significantly more so than our Grave B), it is worth noting the absence of a mouth-covering and of any kind of drinking cup.

Another grave from the same necropolis provides us with an example closer to our Grave A. It was discovered in 1997 and contains a Corinthian exaleiptron; a silver Charon's obol; an iron knife, with a partly preserved handle; a golden mouth-covering; and an iron wire belt.[54] The latter is particularly interesting, as it, once again, appears in conjunction with a blade, but no other types of iron weapons are present.

The graves from Gazoros certainly appear to belong to the same cultural background as Grave A and B. Some (e.g. Grave 20) are notably richer than Grave A and B, but this might be due to the time gap between them and the ensuing differences in burial practices. For instance, the Attic pottery found in some of the rich Gazoros burials from around 475 BCE onwards[55] introduces what may have been perceived as a new type of imported, prestige item in the region. On the other hand, we still find the most common and significant types of metal grave goods (bronze jewellery and accessories, golden or silver mouth-coverings etc.), and they still form a significant part of the burial assemblages, with one noticeable exception: iron weapons seem to have been relatively rare grave goods in the Gazoros region.[56]

In ancient Vergi (modern village of Neos Skopos), more than 100 graves have been discovered north of the ancient settlement, covering a continuous period from the 6th to at least the 2nd century. Unfortunately, we lack precise dating information for almost half of these graves; in addition, from those who are described and dated in detail in excavation reports, only nine have been securely dated to the late 6th / early 5th century.[57]

Here too, imported pottery from both Corinth and Athens makes its appearance, accompanied by a significant number of metal grave goods. Three graves are of particular interest. The first (T41) presents us with a Corinthian exaleiptron; fragmentary bronze

263 (Gazoros), 136-8, 261-2 (Vergi).

50 Samartzidou 1982.

51 Graves 4, 10, 14, 19 and 20, excavated by Poulios in Poulios 1994, 599; 1995, 631-3; also a single unlooted grave excavated by Valla (Valla 1997, 846). Earlier, Samartzidou had excavated fifteen graves "dating from the 6th to the 3rd century BC" (Samartzidou 1982, 329).

52 Samartzidou 1982, 328-9; Poulios 1994, 598-9; 1995, 631-3; Valla 1997, 845-6; 1998, 730; See also Valla 2005 and Kakamanoudis 2018, 347-51 for a general overview of the cemeteries, including earlier structures.

53 Poulios 1995, 632-3.

54 Valla 1997, 846.

55 Poulios 1995, 631-3.

56 Only one spearhead is reported in the 6th century – 3rd century necropolis (Valla 1997, 846). However, more weapons have been discovered in an earlier, 7th century necropolis, to the north of the later burial grounds (Valla 2005, 114-5).

57 Valla 2003, 116-9; see also Kakamanoudis 2018, 343-7.

jewellery including a ring with spiral endpoints; a pair of silver gilded earrings identical to those found in Grave 20 of Gazoros; a golden mouth-covering; and a golden biconical bead.[58] On the scale of rich burial assemblages, this grave lies between our own Grave B and Grave 20, displaying similarities to both, but tending towards the latter with regard to the *selection* of grave goods. The absence of drinking cups and fibulae is notable.

The second grave (T2), found in a separate burial cluster, contained an Attic black-glazed lekythos; a small cup with horizontal bands;[59] an iron belt and knife; a gold mouth-covering; fragments of bronze jewellery including a spiral ring; and a silver stater serving as a Charon's Obol. Somewhat reminiscent of our own Grave A and of the 1997 grave from Gazoros, this assemblage offers us another example of an iron belt associated with an iron knife, further reinforcing this pattern.[60] Other weapons are not encountered (in fact spearheads and swords are, to our knowledge, entirely absent from this Phase of the necropolis).

A Final grave in ancient Vergi (TΛ4) which deserves mention was found in the same cluster as the second one and contained a single child burial, accompanied by an Attic Droop cup and an iron belt with a silver ring clipped onto it.[61] Although relatively poor compared to the other examples presented above, this grave constitutes a very important point of reference for our interpretation of our grave goods (particularly for Grave A), as it demonstrates, firstly, that iron belts could bear additional decoration and aid in displaying personal wealth and, secondly, that they can also be encountered in nonadult graves (without, however, the usual accompanying knife or other weaponry).

As a fourth and final point of comparison, we will also briefly discuss the site of ancient Tragilos, identified near modern-day Aidonochori, west of the Strymon river, on the northern slopes of Mt. Kerdyllion.[62] The site was occupied from before the middle of the 6th century until the early 3rd century, and the burial grounds around and on the slopes of the settlement hill cover most of this timeframe. More than 100 graves (many of which were looted) have been found over multiple excavation sectors. Surveys of the surrounding area indicate that cemeteries also stretched out further to the north, with several hundred more graves having been looted and destroyed during ancient and modern times. A substantial number of the excavated graves (upwards of 70) date to the 6th and the early years of the 5th century. The vast majority are inhumations in cist graves, but a small number of secondary cremations in urns and at least three primary cremations are also encountered.

Grave goods include both locally produced and imported pottery, with locally produced vessels also being employed as funerary urns. Corinthian and Attic pottery is present, including drinking vessels, exaleiptra and aryballoi. Banded cups, similar to those already discussed, are also common. In addition, metal grave goods form a very substantial part of the grave good corpus. They include a wide variety of objects, ranging from bronze, silver, and gold jewellery to bronze *phiale* bowls and to iron tools and weapons.[63] The study of the material from the site, in preparation for publication under the overall direction of Dr. C. Koukouli-Chrysanthaki, has so far revealed a relatively smooth gradient between the poorest and richest graves with regards to metal grave goods. Preliminary investigation of the burial assemblages indicates that the poorest graves may contain little more than a single bronze ring or fibula, whereas the richest typically include large bronze bracelets and biconical beads, intricate spiral hair-coils, and gilded earrings for women; spearheads, knives, swords, and iron belts with spiral endpoints for men; and bronze fibulae, bronze bowls, as well as silver and gold mouth-coverings for both sexes.

We shall refrain from presenting individual graves from Tragilos, pending the completion of the ongoing study and publication of the finds; we shall note, however, that, at the time of writing, the metal artefacts from the Tragilos necropolis seem to correspond almost perfectly to the material culture baseline established by the previously discussed cemeteries. The selection of metal grave goods reflects our observations from other sites, with female graves typically presenting us with the richest and more varied assemblages, while

58 Valla 2003, 116.

59 This type of cup seems to have been among those produced in the Phari workshop (Blondé, Péristeri, & Perreault 1992, 28-30) and other workshops of the region, which recalls the cups found in our two graves.

60 Valla 2003, 118.

61 Valla 2003, 118.

62 For the history of the identification of the site, see Zannis 2014, 129-131 with previous bibliography; for the finds from the cemetery, see below.

63 Koukouli-Chrysanthaki 1971, 417-8; 1972, 530-2; 1973, 452-5; 1978, 297-9; 1983, 123-46; Koukouli-Chrysanthaki & Samartzidou 1985, 270-1; Nikolaidou-Patera 1989 (i), 377-8; 1989 (ii), 483-9; Malamidou 1999, 696; for an overview of the cemeteries see also Kakamanoudis 2018, 351-5.

male graves are more commonly accompanied by iron knives and weapons. Belts are present and are encountered typically in association with knives and other weapons.

Observations and Conclusions

Our overview of the archaeological register clearly demonstrates that metal grave goods are ubiquitous in the Strymon river valley during the late 6th and early 5th centuries. Bronze material (e.g. rings, fibulae) is present even in relatively poor graves, with richer burial contexts providing us with more extensive and varied assemblages of metal objects, including gold and silver artefacts.

Imported pottery from the Thasian region and the central Aegean (Athens, Corinth) has been found in many graves alongside locally produced vessels. However, pending confirmation from archaeometric provenance studies planned for the near future, metal grave goods appear to have been overwhelmingly products of local workshops. They are, therefore, intrinsically linked to local material culture and the insight they can provide is invaluable.

In this cultural context, Graves A and B present us with two relatively rich burial assemblages. Whereas they might fall somewhat short of their richest contemporaries in the overall Strymon region, they compare favourably with the richest graves in nearby Kastas. They also yield very interesting variants of the assemblages typical to rich male and female burials. The presence of the interlinked bracelets in Grave A, which otherwise mirrors perfectly what one would expect of a male inhumation, is unique for the region. Similarly, the iron belt worn by the deceased appears to belong to a relatively rare variant, with significantly smaller and more graceful spiral endpoints than the typical examples from neighbouring sites. It is very likely that such belts were produced in multiple sites and/or workshops.

Grave B is also very important, in that the selection and typology of some of the grave goods distinguishes it from its immediate neighbours. Of note is the pair of silver earrings which, to our knowledge, show no contemporary parallels in the Amphipolis area, or the eastern bank of the Strymon river. The presumed fragmentary silver mouth-covering found in the grave is also noteworthy: during this period, mouth-coverings in nearby Kastas are typically gold, not silver. On the other hand, examples of silver mouth-coverings and near-exact parallels to the earrings in question have been identified west of the river, among the burial assemblages in Tragilos. Such jewellery was probably produced within the region, with individual workshops potentially supplying multiple settlements. That said, given the documented presence of a small number of such earrings even further to the west, networks of diffusion and exchange of metal artefacts are likely to have extended far beyond the Strymon river plain.

Two questions remain to be answered. Firstly: can we discern any symbolic meaning in the selection of metal grave goods, and are social roles and identities (whether biographical to the deceased, or constructed / idealised) communicated through the resulting assemblages? Secondly: Did our two graves belong to local "elites", and which aspects of such a status were emphasised?

Regarding the first question and Grave A, we can, with relative safety, answer in the affirmative. The presence of weapons in the grave does not exist in a vacuum: we have parallels of such assemblages from other nearby sites, and weapons (spearheads and, less frequently, swords) as grave goods in the region form part of a practice going back to the Early Iron Age.[64] This consistency and repetition in funerary ritual argues that weapons were, in fact, symbolically relevant and that they may have served as materialisations of a "warrior" or "hunter" identity in funerary contexts.[65] The typical (but overall rarer) pair of iron belt and knife, which we find in Grave A, sometimes complements these weapons sets; they may have served as an additional part of a warrior's kit or a sign of social status.[66]

64 See e.g. Kastas: 1976, 98; 1977, 42-3; 1981, 25. Gazoros: Valla 2005, 114-5.

65 Whether this social role was "biographical" or "constructed" remains unclear, no less because the poor state of preservation of these iron tools and weapons makes it impossible to discern any traces of actual use on their surface. However, most spearheads discovered in the region, including that of Grave A, still notably preserve traces of wood in their sockets, which would indicate that they were not placed in the grave as isolated, purely symbolic objects, but as part of an (at least theoretically still functional) spear. For "warrior" identities in funerary contexts, see, indicatively, Despoini 2016, 289-90, for a discussion of "warriorhood" as expressed by funerary assemblages in Sindos; Kottaridi 1997, 87; 2009, 89; 2020, 53; for "warrior" graves in Aigai; Whitley 2002, 219-20 and Saripanidi 2017, 93, 96-7, 120-1, with previous bibliography, for a critical examination of "biographical warriorhood" narratives in general and in the Macedonian space in particular, respectively; Kakamanoudis 2018, 439-40 for potentially «constructed» warrior identities in Macedonian Archaic child graves.

66 Although knives could clearly have also served as everyday utility tools, Herodotus (7.74) recounts that Thracians

On the other hand, in female graves, emphasis seems to have been placed more on personal wealth and ornamentation. Typically, silver and gold grave goods are only present in male graves in the form of mouth-coverings, whereas they are encountered in both larger quantities and more varied forms in graves of women. This is also the case with heavy bronze jewellery, including but not limited to bracelets, biconical beads, and fibulae, such as those found in our Grave B. These assemblages shed light on their contemporary ideals of beauty, while also serving as indicators of personal wealth and prestige. Unfortunately, it is, at the time of writing, impossible to determine whether the latter aspects invoke the status and attributes of the deceased herself, or those of her family and/or immediate social circle.

Did, then, the graves presented here belong to "elites"? At the time of writing, we can offer a tentative affirmative answer, while always keeping in mind that the term is relative to the local culture and social structure. Indeed, the Strymon river valley does not yield 6th-century burials rich enough to rival the "princely" graves one can identify in e.g. the burial grounds of sites associated with the Macedonian Kingdom. In addition, there seems to be a relatively smooth gradient of grave assemblage variants between the richest and poorest graves: we see no evidence of a pronounced polarisation between the two extremes, which might have argued in favour of a clearly separate "ruling class".[67]

That said, we have demonstrated that our graves lie, if not at the richest extreme, then certainly near the upper end of the poor-rich gradient. We have also shown that the ritual selection of the grave goods (and metal grave goods in particular) communicate apparently valued social roles and display personal wealth, especially through the inclusion of weaponry (Grave A) and jewellery made of precious metals (Grave B), to which the deceased appear to have enjoyed privileged access. In addition, the deceased were clearly participating in local and supra-local networks of exchange – a conclusion that is further reinforced by the inclusion of imported pottery in the burial assemblages. In this sense, it is not unreasonable to conclude that they were involved in the decision-making process within their local communities and that their activity and agency helped steer the social, political, and economic development of the overall region.

serving in Xerxes' army were armed with "javelins, *pelta* shields, and small daggers".

67 For a thorough discussion of rich burials in the Macedonian Kingdom and proposed interpretations of increased polarisation between rich and poor graves, see Saripanidi 2017, particularly 117-124 with previous litterature. See also Kottaridi 2020, primarily 45-55 for a discussion on Macedonian "princely" burials, an overview of the Aigai necropoleis and a thorough review and interpretation of earlier excavation reports; Chrysostomou & Chrysostomou 2009 and 2012 for a study of visible social hierarchisation in graves from Archontiko based on grave goods.

Bibliography

Alexander, J. 1965, "The spectacle fibulae of southern Europe", *AJA* 69 1, 7-23.

Ardjanliev, P., Chukalev, K., Cvjetićanin, T., Damjanov, M., Krstić, V., Papazovska, A. & Popov, H. (eds), 2018, *100 years of Trebenishte*, Sofia.

Baitinger, H. 2001, *Die Angriffswaffen aus Olympia. Olympische Forschungen* 29, Berlin & New York.

Blondé, F., Péristéri, K. & Perreault, J-Y. 1992, "Un atelier de potier archaïque à Phari (Thasos)", in: F. Blondé, J-Y. Perreault (eds), *Les ateliers de potiers dans le monde grec aux époques géométrique, archaïque et classique. Actes de la Table ronde EFA, 2-3 octobre 1987*, Athens, 11-40.

Boardman, J. & Hayes, J. 1973, *Excavations at Tocra 1963-1965. The Archaic Deposits II and Later Deposits (The British School of Athens Supplementary Volumes 10)*, Oxford.

Bouzek, J. 1974, *Graeco-Macedonian Bronzes: Analysis and Chronology*, Prague.

Carroll, D.L. 1972, "Wire-drawing in antiquity", AJA 76.3, 321-3.

Chavela, K. 2013, "Τοπική γραπτή κεραμική του 6ου π.Χ. από την Τούμπα Θεσσαλονίκης", in: P. Adam-Veleni, E. Kefalidou, D. Tsiafaki (eds), *Κεραμικά εργαστήρια στο βορειοανατολικό Αιγαίο (8ος-αρχές 5ου αι. π.Χ.) Ημερίδα ΑΜΘ 2010*, Thessaloniki, 201-6.

Chrysostomou, A. & Chrysostomou, P. 2001, "Ανασκαφή στη δυτική νεκρόπολη του Αρχοντικού Πέλλας κατά το 2001", *AEMTh* 15, 477-488.

Chrysostomou, A. & Chrysostomou, P. 2006, "Σωστική ανασκαφή στο δυτικό νεκροταφείο Αρχοντικού Πέλλας κατά το 2006", *AEMTh* 20, 703-712.

Chrysostomou, A. & Chrysostomou, P. 2009, "Τα νεκροταφεία του αρχαίου οικισμού στο Αρχοντικό Πέλλας", *AEMTh* 20th-year anniversary volume, 477-490.

Chrysostomou, A. & Chrysostomou, P. 2012, "The 'Goldwearing' Archaic Macedonians from the Western Cemetery of Archontiko, Pella", in: M. Tiverios, P. Nigdelis, P. Adam-Veleni (eds), *Threpteria. Studies on Ancient Macedonia*, Thessaloniki, 490-516.

Despoini, A. 2016, *Σίνδος ΙΙΙ. Το νεκροταφείο. Ανασκαφικές Έρευνες*, Athènes.

Dugas, C. 1928, *Les vases de l'Héraion: Exploration archéologique de Délos X*, Paris.

Kakamanoudis, A. 2018, *Οργάνωση των χώρων ταφής στην αρχαία Μακεδονία. Από την Πρώιμη Εποχή του Σιδήρου έως την Ελληνιστική περίοδο (Unpublished doctoral thesis, University of Thessaloniki)*, viewed 30 May 2019, <https://www.didaktorika.gr/eadd/handle/10442/42957>.

Karamitrou-Mentesidi, G. 2013, "Μαυροπηγή Εορδαίας. Ταφές ανθρώπων και ζώων", *AEMTh* 27, 71-96.

Kilian, K. 1975, *Fibeln in Thessalien von der mykenischen bis zur archaischen Zeit (PBF XIV/2)*, Münich.

Kontogoulidou, E. 2007, "Σωστική ανασκαφή στην αρχαϊκή νεκρόπολη των Αιγών 2007", *AEMTh* 21, 143-148.

Kottaridi, A. 1997, "Βεργίνα 1997", AEMTh 10a, 79-92.

Kottaridi, A. 2009, "Η νεκρόπολη των Αιγών στα αρχαϊκά χρόνια και οι βασιλικές ταφικές συστάδες", AEMTh 20-year anniversary volume, 143-53.

Kottaridi, A. 2020, "The Royal Necropolis of Aigai: walking through the centuries", in A. Kottaridi (ed.), *Macedonian Fragments*, Athens, 45-55.

Koukouli-Chrysanthaki, C. 1971, "Αηδονοχώριον (αρχ. Τράγιλος)", *ArchDelt* 26 (Χρονικά), 417-418.

Koukouli-Chrysanthaki, C. 1972, "Αρχαιότητες και μνημεία Ανατολικής Μακεδονίας", *ArchDelt* 27 (Χρονικά), 520-533.

Koukouli-Chrysanthaki, C. 1973, "Θάσος - Θασιακή Περαία - Νομός Ξάνθης - Νομός Δράμας", *ArchDelt* 28 (Χρονικά), 443-455.

Koukouli-Chrysanthaki, C. 1978, "Τράγιλος (Αηδονοχώρι Ν. Σερρών)", *ArchDelt* 33 (Χρονικά), 297-299.

Koukouli-Chrysanthaki, C. 1979, "Νομός Καβάλας - Νομός Δράμας", *ArchDelt* 34 (Χρονικά), 322-335.

Koukouli-Chrysanthaki, C. 1983, "Οι ανασκαφές στην αρχαία Τράγιλο. Πρώτες γενικές αρχαιολογικές ιστορικές παρατηρήσεις", in *Αρχαία Μακεδονία III*, Thessaloniki, 123-146.

Koukouli-Chrysanthaki, C. 1992, *Πρωτοϊστορική Θάσος: τα Νεκροταφεία του Οικισμού Καστρί*, Athens.

Koukouli-Chrysanthaki, C. 1993, "Η Πρώιμη Εποχή του Σιδήρου στην Ανατολική Μακεδονία", in Αρχαία Μακεδονία *V*, Thessaloniki, 679-697.

Koukouli-Chrysanthaki, C. & Samartzidou, S. 1985, "Αηδονοχώρι (αρχαία Τράγιλος)", *ArchDelt* 40 (Χρονικά), 270-271.

Laffineur, R. 1980, "Collection Paul Canellopoulos (XV)", *BCH* 104 (1), 345-457.

Lazaridis, D. 1964, "Ἀνασκαφαί και ἔρευναι εἰς Ἀμφίπολιν", *Prakt*, 35-40.

Lazaridis, D. 1965, "Ἀνασκαφαί καί ἔρευναί εἰς Ἀμφίπολιν", *Prakt*, 47-52.

Lazaridis, D. 1971, "Ἀνασκαφαί και ἔρευναι εἰς Ἀμφίπολιν", *Prakt*, 50-62.

Lazaridis, D. 1973, "Ἀνασκαφαί και ἔρευναι Ἀμφιπόλεως", *Prakt*, 43-54.

Lazaridis, D. 1974, "Ἀνασκαφαί και ἔρευναι Ἀμφιπόλεως", *Prakt*, 58-64.

Lazaridis, D. 1975, "Ἀνασκαφαί και ἔρευναι Ἀμφιπόλεως", *Prakt*, 61-71.

Lazaridis, D. 1976, "Ἀνασκαφαί καί ἔρευναί Ἀμφιπόλεως", *Prakt*, 88-98.

Lazaridis, D. 1977, "Ἀνασκαφαί και ἔρευναι Ἀμφιπόλεως", *Prakt*, 38-45.

Lazaridis, D. 1978, "Ἀνασκαφαί και ἔρευναι Ἀμφιπόλεως", *Prakt*, 50-58.

Lazaridis, D. 1979, "Ἀνασκαφαί και ἔρευναι Ἀμφιπόλεως", *Prakt*, 71-9.

Lazaridis, D. 1981, "Ἀνασκαφαί και ἔρευναι Ἀμφιπόλεως", *Prakt*, 18-25.

Lazaridis, D. 1982, "Ἀνασκαφαί και ἔρευναι Ἀμφιπόλεως", *Prakt*, 43-51.

Malamidou, D. 1999, "Μεσολακκιά, Αηδονοχώρι", *ArchDelt* 54 (Χρονίκά), 695-696.

Malamidou, D. 2006, "Ανασκαφικά στοιχεία για την παρουσία των Ελλήνων στην περιοχή των εκβολών του Στρυμόνα πριν από την ίδρυση της Αμφίπολης: Εννέα Οδοί καί Άργιλος", *AEMTh* 20, 185-196.

Misailidou-Despotidou, V. 2011, *Χάλκινα κοσμήματα αρχαϊκών χρόνων από τη Μακεδονία*, Thessaloniki.

Neeft, C.W. 2006, "Camarina et la sua Ceramica Corinzia", in: P. Pelagatti, G. Di Stefano, L. De Lachenal (eds), *Camarina: 2600 anni dopo la fondazione. Nuovi studi sulla città e sul territorio*, Ragusa & Rome, 77-107.

Neeft, K. 2020, *The Corinthian Pottery from Argilos (Argilos I)*, Athens.

Nikolaidou-Patera, M. 1989 (i), "Αηδονοχώρι (Αρχαία Τράγιλος)", *ArchDelt* 44 (Χρονικά), 377-378.

Nikolaïdou-Patera, M. 1989 (ii), "Ανασκαφικές έρευνες στις αρχαίες πόλεις Τράγιλο και Φάγρητα", *AEMTh* 3, 483-498.

Oddy, A. 1977, "The production of gold wire in antiquity", *Gold Bulletin* 10, 79-87.

Pabst, S. 2011, "Die grossräumige Ausbreitung der Brillenfiebeln, am Übergang von der Bronze- zur Eisenzeit - Kommunikationswege und soziale Hintergründe", in: U.L. Dietz, A. Jockenhövel (eds), *Bronzen im Spannungsfeld zwischen praktischer Nutzung und symbolischer Bedeutung. Prähistorische Bronzefunde XX.13*, Stuttgart, 199-234.

Payne, H. 1931, *Necrocorinthia: a study of Corinthian art in the Archaic period*, Oxford.

Peristeri, K. 2010, "Σωστική ανασκαφική έρευνα στην περιοχή τύμβου Καστά Αμφίπολης (2009-2010)", *AEMTh* 24, 449-452.

Peristeri, K. 2012, "Ανασκαφή τύμβου Καστά Αμφίπολης 2012", *AEMTh* 26, 531-534.

Peristeri, K. 2013, "Ανασκαφή μνημειακού περιβόλου τύμβου Καστά Αμφίπολης, 2013", *AEMTh* 27, 389-392.

Peristeri, K. 2015, "Ανασκαφική έρευνα τύμβου Καστά Αμφίπολης 2014", *AEMTh* 29, 437-442.

Perron, M. 2013a, "Waveline Household Pottery from Argilos, 6th-5th c. BC: An Overview", in: P. Adam-Veleni, E. Kefalidou, D. Tsiafaki (eds), *Κεραμικά εργαστήρια στο βορειοανατολικό Αιγαίο (8ος-αρχές 5ου αι. π.Χ.) Ημερίδα ΑΜΘ* 2010, Thessaloniki, 133-42

Perron, M. 2013, *La production et la diffusion des céramiques utilitaires de style à bandes à Argilos et dans le Nord de l'Égée aux périodes archaïque et classique (Unpublished doctoral thesis, University of Sorbonne 1 & University of Montreal)*, viewed 25 June 2022, <http://hdl.handle.net/1866/9869>.

Poulios, V. 1994, "Άγιος Χριστόφορος - Γάζωρος", *ArchDelt* 49 (Χρονικά), 598-599.

Poulios, V. 1995, "Άγιος Χριστόφορος - Γάζωρος", *ArchDelt* 50 (Χρονικά), 631-633.

Poulios, V. 2001, "Βέργη - Θάσος - Πίστιρος", *Σερραϊκά Ανάλεκτα* 3, 219-240.

Rhomiopoulou, K. & Kilian-Dirlmeier, I. 1989, "Neue Funde aus der eisenzeitlichen Hügelnekropole von Vergina, griechisch Makedonien", *Prähistorische Zeitschrift*, 86-149.

Risser, M.K. 2001, *Corinthian Conventionalizing Pottery (Corinth 7.5)*, Princeton, New Jersey.

Samartzidou, S. 1982, "Άγιος Χριστόφορος - Γάζωρος - Νέα Ζίχνη", *ArchDelt* 37 (Χρονικά), 328-329.

Samsaris, D. 1976, *Ιστορική γεωγραφία της Ανατολικής Μακεδονίας κατά την Αρχαιότητα* Thessaloniki.

Saripanidi, V. 2017, "Constructing continuities with a "heroic" past: death, feasting and political ideology in the Archaic Macedonian Kingdom", in: I. Lemos, A. Tsingarida (eds), *Constructing Social Identities in Early Iron Age and Archaic Greece. Études d'Archéologie 12*, Brussels, 73-135.

Snodgrass, A. 1964, *Early Greek armour and weapons: from the end of the bronze age to 600 B.C.*, Edinburgh.

Stamatopoulou, V. 2004, *Όπλον: η αργολική ασπίδα και η τεχνολογία της (Unpublished doctoral thesis, University of Thessaloniki)*, viewed 30 May 2019, <https://www.didaktorika.gr/eadd/handle/10442/20347>

Tsimbidou-Auloniti, M. 1989, "Τάφοι κλασικών χρόνων στην Επανωμή", *AEMTh* 3, 319-330.

Valla, M. 1997, "Γάζωρος", *ArchDelt* 52 (Χρονικά), 845-846.

Valla, M. 1998, "Άγιος Χριστόφορος", *ArchDelt* 53 (Χρονικά), 730.

Valla, M. 2003, "Παρατηρήσεις για τη χρήση του χώρου στα όρια της αρχαίας πόλης του Νέου Σκοπού Σερρών", *AEMTh* 17, 111-125.

Valla, M. 2005, "Νεότερα ευρήματα από το νεκροταφείο του Αγίου Χριστοφόρου Σερρών", *AEMTh* 19, 109-118.

Whitley, J. 2002, "Objects with attitude: biographical facts and fallacies in the study of Late Bronze Age and Early Iron Age warrior graves", *CAJ* 12.2, 217-32.

CONTRIBUTORS

Chalazonitis Ioannis
École française d'Athènes, Athens, Greece
CReA-Patrimoine, Université libre de Bruxelles, Belgium
yhal@dhal.com

Georgakopoulou Myrto
STARC, The Cyprus Institute, Cyprus
m.georgakopoulou@ucl.ac.uk

Heymans D. Elon
Faculty of Humanities, University of Amsterdam, Netherlands
elonheymans@gmail.com

Malamidou Dimitria
Ephorate of Antiquities of Serres, Greece
dimitra_malamidou@hotmail.com

Mehofer Mathias
Vienna Institute of Archaeological Science, Human Evolution and Archaeological Sciences, University of Vienna, Austria
mathias.mehofer@univie.ac.at

Nerantzis Nerantzis
Laboratory of Archaeometry, University of the Peloponnese, Greece
CReA-Patrimoine, Université libre de Bruxelles, Belgium
nerantzis.nerantzis@gmail.com

Nodin Sabine
Independent researcher, Perpignan, France
sabinenodin@gmail.com

Oudbashi Omid
Department of Conservation of Cultural and Historical Properties, Art University of Isfahan, Iran
Department of Scientific Research, The Metropolitan Museum of Art, USA
omid.oudbashi@yahoo.com

Papadopoulos Stratis
Ephorate of Antiquities of Kavala-Thasos, Greece
efstratiospa@gmail.com

Photos-Jones Effie
Analytical Services for Art and Archaeology (Ltd), Glasgow, UK
School of Humanities, University of Glasgow, Glasgow, UK
effie.photos-jones@glasgow.ac.uk

Stos-Gale Zofia Anna
Department of Historical Studies, University of Gothenburg, Sweden
zofia@stos-gale.com

Verdan Samuel
University of Lausanne, Switzerland
Swiss School of Archaeology in Athens, Greece
Samuel.Verdan@unil.ch